DYNAMIC STATE VARIABLE MODELS IN ECOLOGY

OXFORD SERIES IN ECOLOGY AND EVOLUTION
Edited by Robert M. May and Paul H. Harvey

DYNAMIC STATE VARIABLE MODELS IN ECOLOGY

Methods and Applications

Colin W. Clark

Marc Mangel

New York   Oxford
OXFORD UNIVERSITY PRESS
2000

Oxford University Press

Oxford   New York
Athens   Auckland   Bangkok   Bogotá   Buenos Aires   Calcutta
Cape Town   Chennai   Dar es Salaam   Delhi   Florence   Hong Kong   Istanbul
Karachi   Kuala Lumpur   Madrid   Melbourne   Mexico City   Mumbai
Nairobi   Paris   São Paulo   Singapore   Taipei   Toronto   Warsaw

and associated companies in
Berlin   Ibadan

Published by Oxford University Press, Inc.
198 Madison Avenue, New York, New York 10016

Library of Congress Cataloging-in-Publication Data
Clark, Colin Whitcomb, 1931–
    Dynamic state variable models in ecology: methods and
    applications / Colin W. Clark and Marc Mangel.
        p.   cm.—(Oxford series in ecology and evolution)
    Includes bibliographical references and index.
    ISBN 0-19-512266-6; ISBN 0-19-512267-4 (pbk)
    1. Animal behavior—Mathematical models.   2. Animal ecology—
Mathematical models.   I. Mangel, Marc.   II. Title.   III. Series.
QL751.65.M3C58   1999
577'.01'5118—dc21   99-12265

1 3 5 7 9 8 6 4 2

Printed in the United States of America
on acid-free paper

Preface

For discussions and comments about the book, we thank Peter Bednekoff, Jane Brockman, Anders Brodin, Tim Collier, Reuven Dukas, Shea Gardner, Paul Hart, Pat Kennedy, Don Ludwig, Barney Luttbeg, Bernie Roitberg, Jay Rosenheim, and Ron Ydenberg.

Many young scientists contributed to the development of the book by participating in courses based on early drafts. We thank at the University of California at Santa Cruz, Edgar Becerra, Melanie Bojanowski, Bret Eldred, Samantha Forde, Eric Danner, Jennifer Brown, Karen Crow, Diana Steller, Matt Kauffman, Kim Heinemeyer, Josh Livni, Diane Thomson, Chris Wilcox, and especially Angie Shelton; at Princeton, Jon Cline, Leila Hadj-Chikh, Tim Kailing, David Smith, Thomas Valqui, and Erika Zavaleta; at the University of British Columbia, Marty Anderies, Chris Fonnesbeck and Edward Gregr; and at Simon Fraser University, Sigal Blay, Julian Christians, and Nick Hughes.

A special vote of thanks goes to Janet Clark, who typeset in $T_{E}X$ many drafts, as well as the final published text.

The original research reported on here has been supported by grants to CWC from the Natural Sciences and Engineering Research Council of Canada, and to MM from the California Sea Grant Program, the National Science Foundation, and the U. S. Department of Agriculture.

We were fortunate to be able to review the final draft on the Big Island of Hawaii, uninterrupted except by omao, yellow tang, and spinner dolphins.

Vancouver, British Columbia   C. W. C.
Santa Cruz, California   M. M.
August 1999

Contents

Introduction

This book expands on an approach to modeling in Ecology that we originally described in our book, M. Mangel and C. W. Clark (1988), *Dynamic modeling in behavioral ecology*, Princeton University Press. The present work is self-contained; there is some overlap with the earlier book, but not much. We have attempted to make the book a self-teaching introduction to the technique and application of dynamic state variable models. Mathematical requirements are quite minimal, but the ability to write computer programs is important. Source code for many of the models discussed in the book is available from the OUP Web site (www.oup-usa.org).

What are dynamic state variable models good for? In their most basic form, they are individual optimization models of behavior. Dynamic state variable models are especially well suited to empirical studies based on field or laboratory data. Each model is designed to reflect the essential biology—with inevitable, well-chosen simplifications. We place much emphasis on obtaining testable predictions from the model and then testing these predictions against the data. We point out that qualitative predictions are often as valuable as quantitative ones, although the latter may also be important.

In the real world, the behavior of any individual is inevitably influenced by the behavior of other organisms, both of the same and of different species. In studying the ecology of populations or communities, we need to develop individual behavioral models that include the density and behavior of other individuals. Such density and frequency dependence can be included in a dynamic state variable model, although this surely makes the model and the computations more complex.

Dynamic state variable models have been used to study human behavior: we describe several such models, remarking in passing on the dangers inherent in assuming that human behavior is determined by biology alone.

A brief comment on terminology: the term *dynamic state variable models* used here is synonymous with *dynamic programming models*, which we used in the previous book. We now prefer the former term as a more accurate description. It is true that these models are solved by the dynamic programming **algorithm**, but we emphasize that other computational algorithms (particularly, forward iteration and Monte Carlo simulation) are also a basic part of the modeling/testing procedure. Given the increasing popularity of these models, we hope that others will also switch to the improved terminology.

DYNAMIC STATE VARIABLE MODELS IN ECOLOGY

1

The Basics

1.1 Introduction: Biology, constraints, and trade-offs

Law (1979) introduced the notion of a Darwinian demon. Such an organism "starts to produce progeny almost immediately after its own birth, producing very large numbers at frequent intervals as it grows older. It experiences no mortality, and its capacity for dispersal and finding mates knows no bounds" (p. 82). We immediately react to such a proposition as ridiculous, and as Law points out, thinking about why Darwinian demons do not exist leads us to useful insights about ecology and evolution.

Darwinian demons do not exist because the world is filled with constraints and trade-offs. Environments cannot provide food resources for limitless numbers of individuals, regardless of how modest the requirement of a single individual may be. Physicochemical processes restrict the rate at which resources can be converted into growth or into offspring. Rapid growth may entail high mortality because the immune system is suppressed or because of exposure to predators. Breeding sites may be limited, leading to conflicts over breeding locations, resulting in injury or death.

This book shows a way of conceptualizing constraints and trade-offs in biology. The methods that we describe allow one to link the physiological states of organisms with the environment via a natural measure of Darwinian fitness. Appropriate applications of the methods suggest new experiments or observations. The methods are accessible to virtually any biologist who has access to a desktop computer and is willing to put in the time learning the new kinds of thinking that are essential for applying the ideas. (These methods are really not all that new—they are related, for example, to classical life-history theory, but they greatly broaden the kinds of adaptations that can be analyzed by that approach.)

Table 1.1 The variables and parameters needed
to characterize a trade-off between starvation,
predation, and reproduction

T	Length of the season (days)
$X(t)$	Energetic reserves at the start of day t
$x_{\text{crit}} = 0$	Critical level of reserves
x_{max}	Maximum possible level of reserves
x_{rep}	Level of reserves needed for reproduction
a_i	Energetic cost of foraging in patch i
m_i	Chance of mortality if patch i is visited
p_i	Probability that food is found in patch i
Y_i	Energetic value of food in patch i
c	Maximum reproductive output in the reproductive patch in a single day

1.2 An example: Formalizing constraints and trade-offs

We begin by developing an artificial model designed to introduce you to
the technique and logic of **dynamic state variable modeling** without getting
involved in intricate biological details. This artificial model will illustrate
such concepts as state variables, constraints, decisions, trade-offs, dynamic
programming, computer implementation, and model experimentation. Later
in this chapter and especially in subsequent chapters, we go into biological
details in a number of specific applications of the technique of dynamic state
variable modeling.

Imagine an organism such as a fish (e.g., Pettersson and Brönmark 1993), a
lizard (e.g., Lichtenbelt 1993), or an invertebrate (e.g. Vadas et al. 1994) that
each day in a season of length T can either forage in one of two feeding patches
or visit a reproductive patch, where it can lay eggs. The feeding patches
provide food resources that are used for both survival and reproduction. At
the end of the season, it dies (if it is an annual organism) or overwinters (if it
is a perennial organism). An overwintering organism has future reproductive
success associated with reproduction in subsequent seasons. At each day in
the current season, we want to predict whether the organism will feed or
reproduce and if it feeds, where it will feed.

To develop such a prediction, we must characterize the organism, the environment, how the two are linked, and how the linkage allows us to make
predictions about its behavior (table 1.1). We characterize the organism by
its **energetic reserves** at the start of day t:

$$X(t) = \text{energetic reserves at the start of day } t \tag{1.1}$$

The reserves are constrained by a **critical value** (which we denote by x_{crit} and

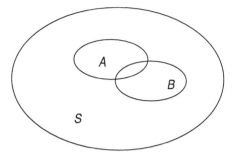

Figure 1.1 The collection of all possible outcomes of an experiment is denoted by S. Smaller collections, which are called events, are contained within S, and the probability of a collection A is the area of A divided by the area of S.

usually set to zero for simplicity), in the sense that the organism dies if the reserves fall to the critical value, and a **maximum value** (denoted by x_{max}) that limits the reserves. Furthermore, reproduction can occur only if reserves exceed a threshold x_{rep}.

Each of the three patches is characterized by costs and risks. We denote the energetic cost (per day) of foraging in patch i by a_i and the daily risk of mortality by m_i, which is the probability that the organism dies if it visits patch i. The feeding patches are further characterized by the probability p_i that food is found in the patch, per day, and by the energetic value Y_i of the food. The reproductive patch is further characterized by the reproductive output c (measured in energetic equivalents) achieved by using that patch.

The organism and the environment are linked through the **expected lifetime reproductive success** of the organism. This allows us to state the problem within the context of a common currency, expected reproductive success. The natural measures of finding food and predation are the probabilities of these events. Thus, before we can evaluate expected reproductive success, a small digression on probability is required.

1.3 A brief digression on probability

In probability theory, we are concerned with the occurrence of "events"—for example, finding food or being killed by a predator. The probability of an event A is denoted by

$$\Pr\{A\} = \text{probability that the event } A \text{ occurs} \qquad (1.2)$$

It is helpful to think of probability in the following way. First, we imagine all the possible outcomes and call this collection of outcomes S. A smaller collection of outcomes A has probability defined as the "area" of A divided by the "area" of S, with "area" suitably defined (fig. 1.1). Courses in probability focus on what the "area of A" means according to the details of the situation. We define

$$Pr\{A\} = \text{probability that the event } A \text{ occurs}$$
$$= \frac{\text{area of } A}{\text{area of } S} \tag{1.3}$$

Continuing to use fig. 1.1 and the definition of probability in eq. 1.3, we see that the probability that at least one of two events A or B occurs is

$$Pr\{A \text{ or } B\} = Pr\{A\} + Pr\{B\} - Pr\{A \text{ and } B\} \tag{1.4}$$

In the future, we will use $Pr\{A, B\}$ for the probability that both A and B occur.

Referring again to fig. 1.1, suppose that we knew that event A occurred. What is the probability that B occurred, given the knowledge about A? This is often called the **conditional probability of B, given** A. If A has occurred, then the collection of all possible outcomes of the experiment is no longer S but must be A. From the definition, eq. 1.3,

$$Pr\{B \text{ occurred, given that } A \text{ occurred}\} = \frac{\text{area common to } A \text{ and } B}{\text{area of } A} \tag{1.5}$$

We use $Pr\{B|A\}$ to denote the conditional probability that B occurs, given that A occurs. Dividing the numerator and denominator of the right-hand side of eq. 1.5 by the area of S and using the new notation,

$$Pr\{B|A\} = Pr\{A, B\}/Pr\{A\} \tag{1.6}$$

By analogy, since A and B are fully interchangeable here, we also require that

$$Pr\{A|B\} = Pr\{A, B\}/Pr\{B\}$$

We define two events as **independent** if knowing that one of them occurred does nothing to change our idea about the probability that the other will occur. Thus if A and B are independent,

$$Pr\{A|B\} = Pr\{A\} \text{ and } Pr\{B|A\} = Pr\{B\}$$

Using the second of these in eq. 1.6, we see that for independent events

$$Pr\{A, B\} = Pr\{A\} Pr\{B\} \tag{1.7}$$

Equation 1.7 is often given as the definition of independent events, but it is actually derived from the definition based on conditional probability.

As described here, the link between the state of the organism and the environment is expected reproductive success. However, at any time $t < T$, the state of the organism at the end of the season (and thus the reproductive success) is a **random variable**—that is, its value is uncertain and is characterized by a probability distribution. In such situations, we define the **expectation** or **expected value** of a random variable as follows. Suppose that Z is a random variable that can take values $z_1, z_2, \ldots, z_K$ according to the probability distribution

$$\Pr\{Z = z_k\} = p_k$$

for which $\sum_{k=1}^{K} p_k = 1$. The expectation of Z is defined as

$$E\{Z\} = \sum_{k=1}^{K} z_k p_k \tag{1.8}$$

Similarly, for any function $g(z)$, we define the expectation of $g(Z)$ by

$$E\{g(Z)\} = \sum_{k=1}^{K} g(z_k) p_k \tag{1.9}$$

Further details on random variables and expectation can be found in Hilborn and Mangel (1997). Now we can continue formalizing the trade-offs in our model.

1.4 The algorithm of stochastic dynamic programming

We now define the function $F(x, t)$ as

$$F(x, t) = \text{maximum expected reproductive success between day } t \tag{1.10}$$
$$\text{and the end of the organism's life, given that } X(t) = x$$

Here, the maximum is taken over possible behaviors (which patch to visit). In biological terms, $F(x, t)$ represents the organism's current "fitness." To begin with, note that we do not know how to obtain the values of the function $F(x, t)$. The rest of this section explains how to calculate $F(x, t)$ by using stochastic dynamic programming.

Our focus is on behavior between $t = 1$ and $t = T$; on day T we summarize all expected future reproductive success by an assumed function $\Phi(X(T))$. Thus, if $X(T) = x$,

$$F(x, T) = \Phi(x) \tag{1.11}$$

Thus, the biology of future reproduction is summarized by $\Phi(x)$.

Next we need to consider days previous to T. Each day, one patch is visited, with fitness consequences. Hence, we let

$$V_i(x, t) = \text{fitness value of visiting patch } i \text{ on day } t, \text{ given} \qquad (1.12)$$
$$\text{that } X(t) = x \text{ and that the organism behaves}$$
$$\text{optimally from day } t + 1 \text{ onward}$$

When $i = 1$ or 2, visiting patch i reduces the state by a_i and may increase it by Y_i, that is, given that $X(t) = x$ and patch $i = 1$ or 2 is visited,

$$X(t+1) = \begin{cases} x - a_i + Y_i & \text{with probability } p_i \\ x - a_i & \text{with probability } (1 - p_i) \end{cases} \qquad (1.13)$$

Each of these has an associated fitness value, which must also take into account avoiding predators. For example, the future expected reproductive success when the animal visits patch i and finds food is $F(x - a_i + Y_i, t + 1)$. There is a similar term when food is not found. Combining these and including the risk of predation shows that the fitness value of visiting patch i, for $i = 1$ or 2, is given by

$$V_i(x, t) = (1 - m_i)\{p_i F(x - a_i + Y_i, t + 1) + (1 - p_i)F(x - a_i, t + 1)\} \quad (1.14)$$

Note that we have implicitly assumed that finding food and avoiding predation are independent events; this allows us to multiply the probability of survival and the expected future fitness. Strictly speaking, eq. 1.14 is really not correct because we've not taken the constraints on the new state into account—that is, the new states must be constrained so that they are never less than $x_{\text{crit}} = 0$ and never larger than x_{max}. This is crucial for computer implementation, and we discuss it in detail in the next section.

Now, consider a visit to a reproductive patch (patch 3). There are three cases to consider:

1. If $x \leq x_{\text{rep}}$, then no reproduction is possible and the fitness of visiting patch 3 is only future expected reproduction:

$$V_3(x, t) = (1 - m_3)F(x - a_3, t + 1) \qquad (1.15a)$$

2. If $x_{\text{rep}} < x \leq x_{\text{rep}} + c$, then reproduction is $(x - x_{\text{rep}})$, and

$$V_3(x, t) = (x - x_{\text{rep}}) + (1 - m_3)F(x_{\text{rep}} - a_3, t + 1) \qquad (1.15b)$$

3. If $x > x_{\text{rep}} + c$, then reproduction is c, and

$$V_3(x, t) = c + (1 - m_3)F(x - c - a_3, t + 1) \qquad (1.15c)$$

Since $F(x, t)$ is the maximum expected reproduction from t onward, given that $X(t) = x$, $F(x,t)$ must be the maximum of the three possible fitness values

$$F(x,t) = \max\{V_1(x,t), V_2(x,t), V_3(x,t)\} \tag{1.16}$$

Equation 1.16 is called a **stochastic dynamic programming equation**. Note that this equation expresses $F(x,t)$ in terms of the functions $V_i(x,t)$ which are themselves expressions involving $F(x',t+1)$ for the various values of $x' = X(t+1)$, from eqs. 1.14 and 1.15. Thus, if we knew the values of $F(x',t+1)$, we could calculate the values of $F(x,t)$ for each x. In other words, we can calculate $F(x,t)$ by working backwards in time from $t = T$, since $F(x,T)$ is given to begin with, by eq. 1.11. We call this procedure **backward iteration**.

To be specific, we use the computer (see section 1.5) to first calculate $F(x, T-1)$ for each value of x, using $F(x',T)$ from eq. 1.11. As we do this, we also find the optimal patch $i = i^*(x, T-1)$, for each x. This is the patch i that gives the maximum fitness value in eq. 1.16. Then we repeat the procedure to find $F(x, T-2)$ and $i^*(x, T-2)$, using the values of $F(x', T-1)$ calculated in the previous step. Thus, each time step ($T-1$, $T-2$, $T-3$, etc.) uses exactly the same method, and this is why it is called "iteration." Modern computers are ideally suited to perform such iterations.

Before describing the computer implementation, we briefly consider an alternative computational method. Why not use the computer to calculate fitness directly for each possible strategy $i(x,t)$, thereby determining the best one? In fact, it is perfectly possible to calculate fitness for any given strategy, using the method of "forward iteration," which we discuss in chapter 2. The problem is that the total number of possible strategies $i(x,t)$ is enormous, so enormous that this approach is not feasible even with a large computer. The backward method, by contrast, is extremely efficient. For example, the entire computation for the patch problem described here takes only a second or two on a typical desktop computer. Backward iteration is fast, efficient, and accurate.

1.5 Computer implementation

We assume that all readers of this book have access to a modern desktop computer. Those who know how to program should simply enter the code given in the next section or visit the Web site http://www.oup-usa.org/sc/0195122674/ or /sc/0195122666/. Our Web site also contains a version of this program written in C++ for whose who prefer that language. Those who do not know how to program should obtain a copy of TRUEBASIC (the language in which the program is written and for which there is a very inexpensive student edition) and enter this program *as soon as possible*; do not spend lots of time learning TRUEBASIC in general, but rather learn it through the program.

The program

The following program implements the dynamic programming equation described in the previous section. Experienced programmers may prefer to skip this and the next subsection and go directly to the following subsection in which the results of the program are explored.

```
dim f(0 to 30, 1 to 20), v(3, 0 to 30, 1 to 20), d(0 to 30, 1 to 20)
dim m(3),p(3),a(3),y(3),c(3)
call parameters
call end_condition
call solve_dpe
sub parameters
let x_crit=0
let x_max=30
let t_max=20
let x_rep=4
print "Parameters are these:"
print " x_crit = ";x_crit;" x_max = ";x_max;" t_max = ";t_max;"
x_rep = ";x_rep
!read the parameters for the three patches
!form of the data is m,p,a,y
print "patch","m","p","a","y"
data .01,.2,1,2
data .05,.5,1,4
for i=1 to 2
read mm,pp,aa,yy
let m(i)=mm
let p(i)=pp
let a(i)=aa
let y(i)=yy
print i,m(i),p(i),a(i),y(i)
next i
!parameters for patch 3
!form of data is m,a,c
data .02,1,4
read mm,aa,cc
let m(3)=mm
let a(3)=aa
let c(3)=cc
end sub
sub end_condition
let acap=60
let x_0=.25*x_max
print
print "x","F(x,T)"
for x=x_crit to x_max
let f(x,t_max) = acap*(x-x_crit)/(x-x_crit+x_0)
print x,f(x,t_max)
next x
!also set the boundary condition that F(x_crit,t)=0
for t=1 to t_max-1
let f(x_crit,t)=0
next t
input prompt "hit any key to go on":k$
```

```
end sub
sub solve_dpe
print
for t=t_max-1 to 1 step -1
print
print "t";tab(10);"x";tab(20);"F(x,t)";tab(30);"d(x,t)";
tab(40);"V(1,x,t)";tab(50);"V(2,x,t)";tab(60);"V(3,x,t)"
for x=x_crit+1 to x_max
for i=1 to 3
!find the fitness values in each patch
if i=1 or i=2 then
!foraging patch
let xp = x - a(i) + y(i)
let xp = min(xp,x_max)
let xpp = x - a(i)
let xpp = max(xpp,x_crit)
let v(i,x,t) = (1-m(i))*(p(i)*f(xp,t+1) +(1-p(i))*f(xpp,t+1))
else
!reproductive patch
if x<x_rep then
let xp = max(x-a(3),x_crit)
let v(3,x,t) = (1-m(3))*f(xp,t+1)
else
let fitness_increment = min(x-x_rep,c(3))
let xp = max(x-a(3)-fitness_increment,x_crit)
let v(3,x,t) = fitness_increment + (1-m(3))*f(xp,t+1)
end if
end if
next i
!find the best patch
let vmax = max(v(1,x,t),v(2,x,t))
let vmax = max(vmax,v(3,x,t))
if vmax=v(1,x,t) then
let d(x,t) = 1
elseif vmax=v(2,x,t) then
let d(x,t) = 2
elseif vmax=v(3,x,t) then
let d(x,t) = 3
end if
!update fitness
let f(x,t) = vmax
!print results
print t;tab(10);x;tab(20);f(x,t);tab(30);d(x,t);tab(40);v(1,x,t);
tab(50);v(2,x,t);tab(60);v(3,x,t)
next x
input prompt "hit any key to go on":k$
!pause 1
next t
end sub
end
```

Explanation of the program

Now we explain the program in detail. To begin, one must tell the computer
to save space in its memory for the various functions that are needed to solve

the dynamic programming equation. This is done in the first two lines:

```
dim f(0 to 30, 1 to 20), v(3, 0 to 30, 1 to 20), d(0 to 30, 1 to 20)
dim m(3),p(3),a(3),y(3),c(3)
```

In the program, we use f(x,t) to denote $F(x,t)$; v(i,x,t) to denote $V_i(x,t)$; d(x,t) to denote the optimal decision; and m(i), p(i), a(i), y(i), and c(i) to denote the parameters m_i, p_i, a_i, Y_i (which is 0 in patch 3), and c_i (which is 0 in patches 1 and 2). Thus, for example, we've told the computer to save 31 spaces in memory for x and 20 spaces in memory for t. Good programming technique involves the use of subroutines, which are little bits of the program that are "called" in the course of running the program. Our program has three subroutines: one sets the parameter values, one sets the end condition eq. 1.11, and one solves the dynamic programming equation eq. 1.16. Since matters must be done in that order, the next lines of the program read

```
call parameters
call end_condition
call solve_dpe
```

Next we explain how the subroutines work. Subroutines are identified writing "sub" followed by the name of the subroutine. The statement "end sub" tells the computer that the end of this subroutine has been reached. The parameter subroutine first specifies x_{crit}, x_{max}, T, and x_{rep}, which are denoted by x_crit, x_max, t_max, and x_rep, respectively:

```
sub parameters
let x_crit=0
let x_max=30
let t_max=20
let x_rep=4
print "Parameters are these:"
print "x_crit = ";x_crit;" x_max = ";x_max;" t_max = ";t_max;"
x_rep = ";x_rep
```

The last two lines print the parameter values on the screen. Next, we read the parameters for the patches. In TRUEBASIC, an exclamation mark (!) denotes a comment or remark in the program, and the computer ignores it when figuring out what to do. We enter the patch parameters by using a DATA statement and a READ statement. When the computer encounters a READ statement, it looks for the first line of data that has not yet been read and puts it into memory.

```
!read the parameters for the three patches
!form of the data is m,p,a,y
print "patch","m","p","a","y"
data .01,.2,1,2
data .05,.5,1,4
```

To read the data, we use a loop for patches 1 and 2. We tell the computer to cycle from $i = 1$ to 2, and for each i, it reads four numbers—which correspond to m_i, p_i, a_i, and Y_i—sets the parameters to these values, and prints them out.

```
for i=1 to 2
read mm,pp,aa,yy
let m(i)=mm
let p(i)=pp
let a(i)=aa
let y(i)=yy
```

```
print i,m(i),p(i),a(i),y(i)
next i
```

The instruction **next i** indicates the end of the *i*-loop.

The parameters for patch 3 are slightly different, since there is no chance of encountering food, but there is a maximum value for reproduction.

```
!parameters for patch 3
!form of data is m,a,c
data .02,1,4
read mm,aa,cc
let m(3)=mm
let a(3)=aa
let c(3)=cc
print "for patch 3:  m = ";m(3);" a_3 = ";a(3);" c = ";c(3)
```

After reading and printing the parameters, this subroutine ends with

```
end sub
```

In this program, we use the following end condition:

$$\Phi(x, T) = \frac{A(x - x_{\text{crit}})}{x - x_{\text{crit}} + x_0} \tag{1.17}$$

where A and x_0 are parameters. Their interpretations are that A is the asymptotic value of $\Phi(x, T)$ when x becomes very large, and $\Phi(x, t) = \frac{1}{2}A$ when $x = x_{\text{crit}} + x_0$. Equation 1.17 corresponds to a future expected fitness that increases with increasing state but does so at a decreasing rate. We need to specify two more parameters, A (called **acap** in the program) and x_0 (called **x_0** in the program)

```
sub end_condition
let acap=60
let x_0=.25*x_max
```

Next, we use a loop, cycling from x_{crit} to x_{max} to compute the end condition given by eq. 1.17. The blank **"print"** line leaves a line of extra space. TRUE-BASIC uses **"let"** to tell the computer to replace the quantity on the left-hand side of the equals sign (=) by the quantity on the right-hand side.

```
print
print "x","F(x,T)"
for x=x_crit to x_max
let f(x,t_max) = acap*(x-x_crit)/(x-x_crit+x_0)
print x,f(x,t_max)
next x
```

Also, we must set $F(x_{\text{crit}}, t) = 0$ for all t, in accord with our assumption that the organism dies if its state falls to x_{crit}.

```
!also set the boundary condition that F(x_crit,t)=0
for t=1 to t_max-1
let f(x_crit,t)=0
next t
```

At the end of the loop, the program asks the user to hit any key to go on; this is done by hitting any key, followed by the RETURN key.

```
input prompt "hit any key to go on":k$
end sub
```

Now we are ready to solve the dynamic programming equation. Recall that we work backwards in time, starting with $t = T - 1$ and going to $t = 1$.

```
sub solve_dpe
print
for t=t_max-1 to 1 step -1
```

At each time, we will cycle over all values of x from $x_{crit} + 1$ to x_{max}. To understand the results, it is helpful to print various kinds of output. Putting items within quotation marks (" ") makes them "strings" that the computer recognizes as labels. The TAB statement moves the label to the appropriate column. We will print $t, x, F(x, t)$, the optimal decision $d(x, t)$, and the three fitness values $V(1, x, t)$, $V(2, x, t)$, and $V(3, x, t)$.

```
print
print "t";tab(10);"x";tab(20);"F(x,t)";tab(30);
    "d(x,t)";tab(40);"V(1,x,t)";tab(50);"V(2,x,t)";tab(60);"V(3,x,t)"
```

Now we cycle over x. For each value of x, we compute the fitness of visiting patch 1, 2, or 3.

```
for x=x_crit+1 to x_max
for i=1 to 3
!find the fitness values in each patch
```

If $i = 1$ or 2, the animal visits a foraging patch. Now we take into account the constraints on the state variable. In particular, we let $x' = \min(x - a_i + Y_i, x_{max})$; we denote x' by xp. We let $x'' = \max(x - a_i, x_{crit})$; we denote x'' by xpp. The fitness value of visiting patch 1 or 2 is given by eq. 1.14, which becomes

```
if i=1 or i=2 then
!foraging patch
let xp = x - a(i) + y(i)
let xp = min(xp,x_max)
let xpp = x - a(i)
let xpp = max(xpp,x_crit)
let v(i,x,t) = (1-m(i))*(p(i)*f(xp,t+1) + (1-p(i))*f(xpp,t+1))
```

Otherwise, the organism visits the reproductive patch. If $x < x_{rep}$, then it receives no immediate increment in fitness but still has future fitness determined by the new state; otherwise, it receives a fitness increment that is the minimum of c and $x - x_{rep}$. In this case, the new state is $x' = \max\{x - a_3 - \min(x - x_{rep}, c), x_{crit}\}$, so that

```
else
!reproductive patch
if x<x_rep then
let xp = max(x-a(3),x_crit)
let v(3,x,t) = (1-m(3))*f(xp,t+1)
else
let fitness_increment = min(x-x_rep,c(3))
let xp = max(x-a(3)-fitness_increment,x_crit)
let v(3,x,t) = fitness_increment + (1-m(3))*f(xp,t+1)
end if
end if
next i
```

We find the best patch by comparing $V_1(x, t)$ and $V_2(x, t)$ and call the larger of these V_{max} (denoted by vmax). Then we compare V_{max} with $V_3(x, t)$ and choose the optimal decision by comparing V_{max} with the three fitness values.

```
!find the best patch
let vmax = max(v(1,x,t),v(2,x,t))
let vmax = max(vmax,v(3,x,t))
if vmax=v(1,x,t) then
let d(x,t) = 1
elseif vmax=v(2,x,t) then
let d(x,t) = 2
elseif vmax=v(3,x,t) then
let d(x,t) = 3
end if
```

Next, we update fitness according to eq. 1.16, print the results, and continue to cycle over x.

```
!update fitness
let f(x,t) = vmax
!print results
print t;tab(10);x;tab(20);f(x,t);tab(30);d(x,t);tab(40);
v(1,x,t);tab(50);v(2,x,t);tab(60);v(3,x,t)
next x
```

Then, either we stop the program until a key is hit, or we pause for 1 second and then tell the computer that this subroutine is ended:

```
input prompt "hit any key to go on":k$
!pause 1
next t
end sub
```

Finally, we tell the computer that the program is completed:

```
end
```

For neophytes, this may seem daunting. But do not be discouraged; hundreds of biologists have learned how to apply dynamic state variable models, beginning with little or no knowledge in programming and using this kind of model and associated program as a template for their own work.

Explorations with the program

For the parameters chosen, patch 1 is the safer but less productive foraging patch, patch 2 is the riskier but more productive foraging patch, and the mortality risk in the reproductive patch is intermediate between the two. The output from this program (table 1.2) will show the time, state, maximum expected reproductive success, optimal behavior, and fitness of each of the three options. The latter information allows us to assess how important it is for the organism to behave optimally. For example, at $x = 21$ in table 1.2, the fitness "cost" of visiting patch 2 or patch 3, rather than patch 1, is less than 1%. On the other hand, at $x = 4$, the fitness cost of visiting patch 1 or 3, rather than the optimal patch, is about 10%. Thus, we expect only weak selection on behavior for $x = 21$ but strong selection for behavior when $x = 4$. McNamara and Houston (1986) called such values the canonical cost of a one-time error, assuming that subsequently the organism behaves optimally.

Table 1.2 Sample output from the program

t	x	$F(x,t)$	$d(x,t)$	$V(1,x,t)$	$V(2,x,t)$	$V(3,x,t)$
15	1	11.0724	2	3.471	11.0724	0
15	2	17.1434	2	12.7917	17.1434	10.7405
15	3	20.7761	2	18.1143	20.7761	16.7031
15	4	23.19	2	21.4228	23.19	20.3507
15	5	24.9828	2	23.7474	24.9828	22.8442
15	6	26.3953	2	25.4593	26.3953	24.6291
15	7	27.6269	2	26.941	27.6269	26.1612
15	8	28.7207	2	28.2462	28.7207	27.494
15	9	29.7574	2	29.4241	29.7574	28.494
15	10	•30.7614	2	30.5233	30.7614	29.494
15	11	31.666	2	31.5273	31.666	30.494
15	12	32.5415	2	32.4645	32.5415	31.494
15	13	33.4396	2	33.3921	33.4396	32.6993
15	14	34.3445	2	34.3319	34.3445	33.8284
15	15	35.2105	1	35.2105	35.1743	34.8374
15	16	36.1104	1	36.1104	35.9884	35.7614
15	17	37.0262	1	37.0262	36.8171	36.6949
15	18	37.9346	1	37.9346	37.6402	37.6372
15	19	38.7849	1	38.7849	38.4716	38.4945
15	20	39.5855	1	39.5855	39.2446	39.3768
15	21	40.3981	1	40.3981	40.0015	40.2963
15	22	41.221	3	41.1842	40.7426	41.221
15	23	42.0756	3	42.0356	41.5607	42.0756
15	24	42.8729	3	42.8329	42.3372	42.8729
15	25	43.6633	3	43.6233	43.0781	43.6633
15	26	44.4366	3	44.3966	43.798	44.4366
15	27	45.2973	3	45.2181	44.5807	45.2973
15	28	46.0948	3	46.0156	44.9706	46.0948
15	29	46.8661	3	46.7869	45.3377	46.8661
15	30	47.6218	3	47.3765	45.6913	47.6218

Note additionally that although $x_{\text{rep}} = 4$, the optimal trade-off involves "delaying" reproduction and building up the state if it is relatively low.

The output exemplified in table 1.2 divides the state space into three regions (fig. 1.2). When x is sufficiently small, patch 2 is visited. As the state increases, either patch 1 is visited (if time is sufficiently close to the time horizon), or patch 3 is visited. The exact nature of the boundaries is determined by the parameter values, and the best way to understand what forces shape the boundaries is through computer "experiments."

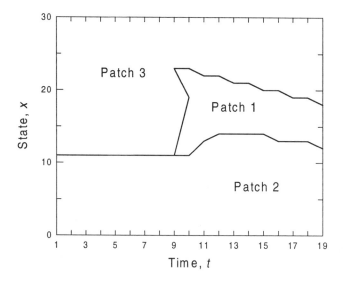

Figure 1.2 The solution of the state variable model divides the time-state plane into regions according to the predicted optimal behavior.

Possible experiments with the program

By changing parameter values, we can experiment with the program and use the output to improve intuition about what is driving the results, as well as to check the program itself. The following are some suggestions:

1. Increase m_2 to 0.05. In this case, patch 1 is now equally risky as patch 2 but not as productive. It will never be optimal to visit patch 1.
2. Set $c = 0$. In this case, there is no reproductive payoff from the reproductive patch; it will never be optimal to visit patch 3. (Here, all reproduction occurs at time T or later.)
3. Set $A = 0$. In this case, there is no future reproduction; the reproductive patch will be visited at lower state values.

These predictions are straightforward because of the nature of the trade-offs. Some others, which we suggest that you explore, are not nearly as clear:

4. Attribute a fitness increment only if the organism survives the period in the reproductive patch. This involves changing the fitness value of visiting patch 3 to

$$
\begin{aligned}
V_3(x,t) = (1 - m_3)\{&\min(x - x_{\text{rep}}, c) \\
&+ F(x - a_3 - \min(x - x_{\text{rep}}, c), t + 1)\}
\end{aligned}
\tag{1.18}
$$

and the associated line in the program to

```
let v(3,x,t) = (1-m(3))*(fitness_increment + f(xp,t+1))
```

We encourage you to try this.

5. Varying the other patch parameters. Any of m_i, a_i, p_i, Y_i, or c can be varied, and we encourage you to think first about the possible effect and then to try the experiment. For example, what happens as m_2 increases? What happens if $p_2 Y_2$ is held constant, but the parameters p_2 and Y_2 are varied? Can you interpret the results (and make predictions) in terms of trade-offs between survival and reproduction? (In chapter 3 we will have a broader discussion of using models to make predictions.)

1.6 Evaluating the fitness of alternative strategies

It is appropriate to ask if behaving optimally is important. For example, we might wonder how an animal that randomly picked patches would survive compared to an animal that visited patches according to the optimal rules. Suppose that $F_r(x, t)$ is the expected reproductive success of an individual that randomly visits patches. We can calculate $F_r(x, t)$ by using a modified dynamic programming equation. The end condition is the same since the future fitness is a function only of state at T. The dynamic programming equation changes, however. Assume that each patch is picked with probability $1/3$. Then instead of eq. 1.16

$$F_r(x, t) = \frac{1}{3}\{V_1(x, t) + V_2(x, t) + V_3(x, t)\} \qquad (1.19)$$

where now the $V_i(x, t)$ have $F_r(x', t + 1)$ on the right-hand side. The results (fig. 1.3) show that there is considerable fitness advantage to following the optimal pattern of patch choice. In this figure, we plot $F(x, 1)$ and $F_r(x, 1)$ as functions of x. If the initial state is close to the critical value, an animal that follows the optimal behavior has expected reproductive success about 10 times greater than an animal that follows the random behavior. If the initial state is $x_{\max}$, the difference is still about 30%.

A second alternative behavior might be the following. Suppose that $c = 0$, so that reproductive success is attained only at the end. Clearly, it is better to be bigger at the end than to be smaller, so why not visit patch 2 (the more profitable but riskier patch) in every period? To answer this question, we set $c = 0$ in the program and let i range from 1 to 2, rather than from 1 to 3. The results (fig. 1.4) show that if the state is far from the critical value, there is considerable advantage to protecting the asset (high state) by visiting the safer patch. When the state is close to the critical value, the advantage is smaller because the optimal behavior itself involves visiting the riskier patch (patch 2).

Finally, it is often proposed that animals use "rules of thumb" as approximations for optimal behavior. For example, one rule of thumb might be to

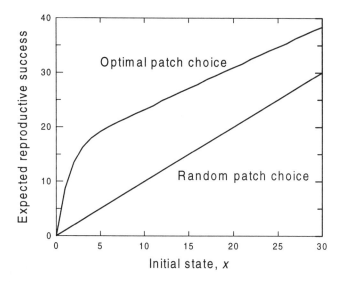

Figure 1.3 That it pays to behave optimally is shown by comparing the expected reproductive success $F(x, 1)$, assuming that the organism behaves optimally, with the expected reproductive success $F_r(x, 1)$ obtained by random choice of patches.

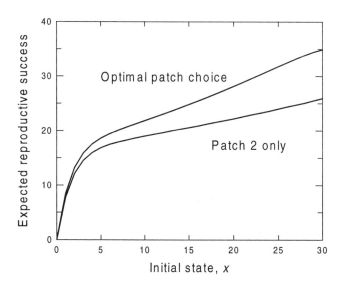

Figure 1.4 Fitness for the optimal patch choice compared with fitness if the forager always visits patch 2, for the case in which there is only terminal fitness (i.e., $c = 0$).

introduce a threshold x_{th} and choose the riskier patch whenever the state is below this threshold, and choose the safe patch when the state equals or exceeds the threshold.

$$\text{Patch choice} = \begin{cases} 2 & \text{if } x < x_{\text{th}} \\ 1 & \text{if } x \geq x_{\text{th}} \end{cases} \tag{1.20}$$

We encourage you to modify the program and compare the expected reproductive success of animals following the optimal rules and the threshold rule of thumb.

1.7 Developing a dynamic state variable model

Now we explain how, in general, to construct a dynamic state variable model for a given set of observations, after which we develop a model for diet choice in the three-spined stickleback. In subsequent chapters we discuss several more complicated examples. We also show how to derive testable predictions from a dynamic state variable model. The question of comparing predictions with data is discussed in chapter 3.

The first step in building a dynamic state variable model is to know the biology needed to inspire the model and to obtain the data needed to parametrize the model. To some extent this is a chicken-and-egg situation. How do you know what data to collect before you have a model in mind? And how do you make up a model if you don't have any data? Ideally, model building and data collection go hand in hand. But let us assume that some interesting data have already been obtained (e.g., that diet choice in three-spined sticklebacks does not follow classic rate-maximizing behavior). We have decided that a dynamic state variable model will help us to understand the data better. As we develop the model, we may decide that more or different data will be needed.

Basic steps

The following sequence of steps should be followed in developing a dynamic state variable model:

1. Specify the basic time interval and time horizon.
2. Specify the state variables and constraints.
3. Specify the decision variables.
4. Specify the state dynamics, including stochastic aspects.
5. Specify the characterization of fitness and the terminal fitness function.
6. On the basis of steps 1–5, write the dynamic programming equation.
7. Parametrize the model by specifying parameter values and functional forms.
8. Write the computer code for the model, and perform test runs.

Upon completing these steps, the model can be used to generate predictions.

In going through these eight steps, you will almost invariably find that certain modeling decisions made at an earlier stage will need to be modified or made more explicit as you proceed through later stages. You will also have to decide between including or omitting various biological details. We recommend simplicity over complexity at this stage, especially during the initial development of a dynamic state variable model. It is a common error to try to include in the model "everything you know" about the system. Avoid this temptation. Good science derives maximum explanation from minimum hypotheses. Where to simplify by omitting detail is always a matter of delicacy and style. Our modeling philosophy is not to include details unless they contribute to the explanatory or predictive power of the model. Of course, at first, you may not know how important certain details may be. If so, comparing the predictions of different versions of the model will be revealing.

As you will discover, the process of going through the eight steps forces you to think about what are likely to be the most important factors that affect the system that you are modeling. This is one of the major payoffs from the modeling process.

Steps 1–5 specify the structure of your model. At step 6, you encapsulate these components in the dynamic programming equation. At this stage, you will probably discover that further detailed modeling decisions are required. Being able to write the dynamic programming equation with absolute confidence is the touchstone for ensuring that your dynamic state variable model is compete and logically consistent. This may take some time, with false starts and blunders, but persevere. Some individuals tend to skip this essential stage and go from the specification of state variables, etc., directly to the computer keyboard. This is a serious tactical mistake, which these individuals pay for many times over, in terms of trying to find computer bugs and (even more devastating) lack of a complete understanding of the assumptions underlying the model. You may not be accustomed to writing mathematical equations, but taking the time to learn this skill will reward your efforts—we guarantee it.

More on the basic steps

Although the basic steps in developing a dynamic state variable model are common to all such models, the details of specifying model components are extremely flexible. Each separate model involves unique features. We still often encounter new twists when developing a new model.

Basic time period
In choosing the duration of the basic time period (i.e., the difference between t and $t+1$), the main consideration is that each time period should involve at most one decision of each type made by the organism. For example, in a model of search and foraging behavior, where the question is whether an item of food encountered should be accepted or rejected (or perhaps whether it should be eaten immediately or cached), the time period should be brief enough that the likelihood of encountering more than a single food item per period is negligible.

Typically, a period of a few seconds or minutes could be appropriate here (e.g., Mangel 1989; Mangel and Clark 1988). For a microbehavioral model like this, the overall span of time considered would be fairly brief, perhaps one or a few days. For example, a 14-h day broken into 5-min periods would imply $T = 168$ periods per day. (How to link successive days and nights is discussed in the next chapter.) Houston and McNamara (1993) modeled daily foraging routines of song birds in winter, using models of this kind.

For a model covering several months or years, a 5-min basic time period would obviously require a very large number of time steps. It's a matter of appropriate scales—microbehavior should usually be modeled over a relatively brief time span, and a coarser decision submodel should be used in modeling over a longer time span. For example, Clark and Ekman (1995) modeled wintering foraging strategies of willow tits (*Parus montanus*) in Sweden over the 180-day winter; Bull et al. (1996) modeled the overwinter behavior of juvenile Atlantic salmon (*Salmo salar*). In both cases, the basic time period was one day; behaviors were which foraging habitat to use for the given day or how much of the day to forage (we discuss this work in chapter 5). The models yielded useful predictions, even though a shorter decision period of one hour or less might have been more realistic.

The basic time interval does not need to be the same for the entire model. For example, in modeling foraging or oviposition behavior over several days, 5-min periods might be used during the day, and a single period for the night. No decisions are made overnight, but daytime foraging strategies must try to obtain sufficient energy reserves to meet night metabolic costs (Houston and McNamara 1993; Clark and Ekman 1995). Then the dynamic programming algorithm has two cases, day and night, and this is treated by sequential coupling (see chapter 2).

The choice of a basic time period is determined to some extent by other components of the model. Generally all of the model's components are linked, and choices made about modeling one component will often have implications for the rest. Thus going step-by-step through the list of components is usually an iterative process of successive refinement.

Time horizon

Computer solution of the dynamic programming equation requires that one begin at some terminal time or time horizon T and iterate backward in time. Usually the value of T is determined by the total time interval over which one wishes to model the state and decisions (together with the length of one basic time period).

But what should T be for a model covering an entire life span? Whether or not the species in question undergoes senescence, the probability of living beyond a certain age is negligible, and T can be chosen to cover this maximum predicted life span.

Another possibility is that the terminal time is random. For example, an insect may die on the first day of frost. This is handled by specifying T as the last possible day of first frost. If $w(t)$ is the probability that first frost occurs at time period $t \leq T$, then the conditional probability $w_f(t)$ that first frost occurs at time t, given that it has not occurred before t, is expressed as (this is an example of conditional probability, eq. 1.6)

$$w_f(t) = \text{Pr}\{\text{first frost occurs on } t, \text{ given that it has not occurred yet}\}$$

$$= \frac{\text{Pr}\{\text{first frost occurs on } t, \text{ and it has not occurred yet}\}}{\text{Pr}\{\text{first frost has not occurred yet}\}}$$

$$= \frac{w(t)}{\sum_{s=t}^{T} w(s)} \tag{1.21}$$

(so, for example, $w_f(T) = 1$).

State variables
One can readily imagine that the choice of state variables can vary widely. The organism's physiological state (e.g., fat or nutrient reserves, body mass, egg complement, and oxygen debt) is often represented in terms of state variables. Environmental variables (e.g., temperature, wind strength and direction, geographical location of the organism, and density of prey or hosts) may also be state variables. Various kinds of information (e.g., estimated predation risk or prey abundance) can also be treated as state variables.

Some of these state variables are affected by behaviors of the organism, whereas others—for example environmental state variables—may not be. The current values of both kinds of state variables also influence current behavior. The state variables included in a given model are all those variables that change over time and that influence or are influenced by the organism's activities (as far as the model is concerned). As discussed later, the organism's fitness function $F(x, y, \ldots, t)$ is a function of these state variables, or to be more precise, of the current values $x = X(t)$, $y = Y(t)$, $\ldots$ of the state variables. Similarly, current decisions D depend on the state variables, $D = D(x, y, \ldots, t)$. In general, both fitness and decisions also depend on the time variable t.

It is this feature, whereby both current fitness and current actions are considered to depend on and affect dynamic state variables, that characterizes the dynamic state variable modeling paradigm and separates this approach from traditional life-history and behavioral modeling. For example, life-history traits (such as age of maturity and number of offspring) are sometimes considered to depend on an external environmental state E. This is called **phenotypic plasticity**, and the functional relationship between E and the given life-history trait is called a **reaction norm** (Roff 1992; Stearns 1992). Thus, dynamic state variable models are a generalization of traditional life-history theory.

In a dynamic state variable model, it might appear that current decisions can depend only on present conditions and that neither involve past experience nor anticipate possible future events. Quite the opposite is true, however: dynamic state variable models almost automatically include both past and future considerations. First, past history is incorporated in the current values of the state variables. For example, an animal's current body mass reflects its past history of growth. Likewise, a migratory bird's current location reflects its past migratory behavior. Second, the dynamic programming algorithm considers both current reproduction (if any) and future reproductive success. If an animal expects that future environmental conditions will change, the dynamic model takes this knowledge into account. As one example, many dynamic state variable models predict that animals will change their foraging strategies before the change of seasons and build up fat reserves for future breeding or as a preparation for winter conditions. No other modeling approach has this feature.

Constraints, or limits, on the range of values of a state variable are usually straightforward—an animal can eat only so much or can grow only to a certain size. Finite limits on the values of variables are required in any computer realization. Many analytic models do not include such constraints, which may be hard to deal with analytically (see Mangel 1992), and spurious predictions are possible. One important consideration in selecting state variables is that computational complexity increases exponentially with the number of state variables. Dynamic programming models that have more than four or five continuous state variables quickly swamp computer time and memory capacity. Our advice is to minimize the number of state variables.

Decision variables

We use the word *decision* to indicate behavior when the organism has a number of possible behaviors; cognitive choice is not assumed. Selecting decision variables is often quite straightforward, given that the objective of modeling is often to explain why animals make certain decisions, rather than others that could be made. Some examples are the number of eggs a parasitic wasp lays on a given host (Mangel 1987); a nesting bird's decision to stay at the nest, hunt for food, or desert its mate (Kelly and Kennedy 1993); a forager's decision to forage with a group or act as a sentinel (Bednekoff 1997); a migratory bird's decision to initiate a migratory flight or to continue feeding (Clark and Butler 1999); and the decision of a fish to attack a given prey or bypass it (Hart and Gill 1992a).

Some decision variables may naturally be discrete, that is, they may have only a finite number of possible values; all of the examples listed above are of this type. Other decision variables may be continuous, such as the fraction of new resources to allocate to growth and to reproduction, when both occur simultaneously. In the computer realization, even continuous decision variables have to be discretized—more on this later. For continuous decision variables, the range of possible values is usually self-evident. For example, if a_1 and a_2 denote proportional allocation of time between two activities (e.g., foraging or searching for a mate), we require that $0 \le a_1 \le 1$ and $0 \le a_2 \le 1 - a_1$ because total portions must sum at most to 1.

The set of feasible decisions may also be time-dependent. For example, diurnal predators cannot search for prey at night; most temperate zone animals cannot breed in winter. As indicated in chapter 2, such temporal variation can be included by sequentially coupling separate submodels for each different temporal circumstance.

The outcome of a given decision may involve stochastic aspects. For example, during a migratory flight, a bird may encounter head or tail winds, which will affect the duration of the flight and the energy reserves that remain at the end of the flight. Thus, decisions, state dynamics, and stochastics are often closely intertwined.

State dynamics and stochastics

The change in an organism's state from period t to $t + 1$ generally depends on the current states $x, y, \ldots$, and on the decisions taken in period t. These changes may involve stochastic components (luck in finding food, size of food item found, weather-related metabolic costs, and predation events). The clue to including stochastic aspects in a dynamic state variable model is the belief that some particular source of randomness may be important in determining observed behavior. Intuition is seldom a reliable guide to understanding randomness and uncertainty (Bernstein 1996), but a carefully thought out dynamic model can often help by sharpening and supporting intuition—or perhaps by contradicting and altering intuitive ideas.

The actual form of the probability distribution used in modeling stochasticity is usually much less important than the decision to include stochasticity in the first place. Of course, there may be theoretical or empirical reasons to use a particular distribution. But even an ad hoc (for example, discrete) probability distribution is often useful. The sensitivity of predictions to the variance of the assumed distribution can easily be determined and provides a measure of the stochastic component's importance.

In a model including stochasticity, first one must decide what to assume about the animal's reaction to random events. Which comes first within a single time period, the decision or the event? In the patch-selection model described previously, for example, first, the forager decides on which patch to visit. Once in the patch, it may or may not be killed by a predator. If not killed, in patch 1 or 2, it may or may not find food. Hence the fitness value of visiting patch 1 or 2 has the form of eq. 1.14. The logic is as follows: first, the organism considers each possible patch i, resulting in a certain value for expected fitness if it chooses that patch. Then it selects the patch with the largest fitness.

In a model of oviposition behavior (Mangel and Clark 1988), it is necessary to assume that the decision (clutch size) is made after the random event (encountering a host) occurs. In this case, the dynamic programming equation has a different form (see chapter 4)

$$F(x, t) = \sum_{j=0}^{J} p_j \max_{c \leq x} \{w_j(c) + (1 - m_j)F(x - c, t + 1)\} \tag{1.22}$$

Here, x is the number of eggs that the insect harbors at the start of period t, $w_j(c)$ is the fitness increment from a clutch c on a host of type j, and m_j is mortality during oviposition on a host of type j. A host of type j is first encountered (type 0 means no host) with probability p_j, after which the decision to lay a clutch of size c is made. It would even be possible to have a decision-event-decision sequence in each time period. For example, a wasp might first select a search area, then encounter a host, and finally lay a clutch.

Two additional important aspects of stochasticity are behavioral plasticity and learning (Dukas 1998). To adjust its behavior to varying environmental conditions, an organism must be capable of both sensing different conditions, and processing the sensory information. The sensory data available to the organism may provide only partial information about the current state of the environment. For example, a forager may be aware that certain patches are normally more productive than others, but productivity may vary randomly over time. Only by sampling the patches can the forager learn the current conditions, and sampling error may mask the actual state of the patch.

Natural selection has led to the evolution of highly developed senses and rapid information-processing ability, but even a perfect information system cannot instantly determine the current state of a fluctuating environment. Modeling the dynamics of uncertainty is a difficult challenge; in chapter 11 we discuss this question in greater detail.

Fitness characterization
It is important to be absolutely explicit about the definition of the fitness function $F(x, t)$ you will use, because the form of the dynamic programming equation follows directly from this definition. Frequently $F(x, t)$ is defined as expected reproductive success from period t to the end of the individual's life, given that $X(t) = x$. (The case of several state variables $x, y, \ldots$ is similar.) In life-history theory this measure is usually denoted by R_0 (Roff 1992; Stearns 1992); approaches using the Malthusian parameter r are discussed in chapter 12. In this book we use mainly R_0, although in some models a simpler fitness characterization such as the probability of surviving a winter may be appropriate.

Note that fitness $F(x, t)$ is defined in terms of present and future reproduction. All past reproduction is ignored, and current decisions are affected only by current and future reproductive opportunities. It is important not to make the mistake of including past reproduction in the fitness function. Failure to adhere to this rule in analyzing parental care has been termed the "Concorde fallacy" (Dawkins and Carlisle 1976). It is possible to misunderstand this rule, however. For a parent currently feeding or otherwise caring for offspring, the presence of these offspring certainly affects current behavior and future fitness. Hence, the offsprings' states become part of the parent's state variables. Then, the parent's fitness function $F(x, y, t)$ depends on the parent's current state x and the state y of offspring being cared for. Once offspring become independent, they no longer influence the parent's fitness function. Clark and

Ydenberg (1990a,b) discussed parental care and fledging of seabirds using this method.

Terminal fitness
If the time horizon T is the last possible period in which the organism can survive, the terminal fitness function is expressed as

$$F(x, T) = 0 \qquad (1.23)$$

In other models, where only a part of the organism's life span is under consideration, the terminal condition becomes

$$F(x, T) = \Phi(x) \qquad (1.24)$$

where $\Phi(x)$ is the expected future lifetime reproduction from time T to the end of the organism's life, given that $X(T) = x$. Specification of the terminal fitness function $\Phi(x)$ (except when it is zero) is part of the model construction process. There are various ways in which $\Phi(x)$ could be quantified. First, at least in principle, $\Phi(x)$ could be measured experimentally. This has often been done for semelparous organisms that reproduce once and then die. In this case, it is often found that $\Phi(x)$ is approximately linear above some threshold level x_{th}, at least in the event that the state variable x is related to body size or fat reserves.

For a migratory species with fitness function $F(x, y, t)$ that depends on body size x and geographical location y, one might have

$$F(x, y, T) = \begin{cases} \Phi(x) & \text{if } y = y_b \\ 0 & \text{otherwise} \end{cases} \qquad (1.25)$$

where location y_b is the breeding ground. This specifies that no breeding occurs unless the organism reaches the breeding ground. An example of this type is discussed in chapter 6.

For iteroparous organisms that can live to reproduce again later in life, the specification of terminal fitness is more difficult. However, since $\Phi(x)$ represents future fitness at time T, one can imagine calculating $\Phi(x)$ from another model in which the first period coincides with period $T + 1$ of the given model. This is an example of sequential coupling, which is discussed in detail in the next chapter.

A final comment on terminal fitness: when the dynamic programming equation has a **stationary solution**, so that the decision function is independent of time t for $t \ll T$, the stationary solution is also independent of $\Phi(x)$. As explained by McNamara and Houston (1982), one would not expect behavior at time t to be much influenced by events that will occur in the distant future T, where $t \ll T$. An example is the fattening strategies of small birds in winter, as discussed in chapter 5.

The dynamic programming equation

Having completed steps 1 through 5, the next step is to derive the dynamic programming equation. Indeed, the very process of writing the dynamic programming equation serves as a check that you have completely and unambiguously specified all the components of your model. Obtaining the dynamic programming equation is not always entirely routine. Each new model seems to involve some novel twist. Don't be satisfied with guesswork. Your dynamic programming equation must follow logically from your stated assumptions. As a check, try deriving the equation twice, a week apart. If you are working with a colleague, compare independent derivations, and convince each other. Above all, do not try to skip this crucial step by prematurely attempting to write the computer code for your model.

Parameters and functions

In an ideal world, there would be sufficient data to estimate the model parameters and to fit functions. In this ideal world there would also be little use for models (or thinking in general). It is a mistake to think that a model should be attempted only after the biological system is fully understood. Indeed, if the system is fully understood, why should anyone bother working on it? On the contrary, building a model can help to reveal what else needs to be learned about the system.

Then, what should one do when some parameters or functional forms of the model are not known? Often, reasonable guesses at parameter values can be obtained by interpolating or extrapolating from data for related organisms by using allometric relationships. Basal metabolic rate is an example. Sometimes simple preliminary models can be used to generate reasonable functional forms. For example, in modeling diel vertical migration, Clark and Levy (1988) needed a functional relationship between light intensity in the water and the foraging success rate of predators. They used a cylindrical search-volume model to generate this relationship. The search model also revealed the role of other basic ecological parameters such as swim speed, prey density, and handling time in the phenomenon of vertical migration and led to a new adaptive explanation of the observed behavior of certain planktivores. When in doubt build a model—and keep it simple!

Finally, what if certain parameter values are simply unknown? The only recourse is to use a reasonable guess and to test the sensitivity of model predictions to the unknown value or values. There is even scope here for a bit of "reverse" prediction and model testing; that is, one can tune a certain unknown parameter until the model generates predictions that agree with observations and then predict that this must be the value of the parameter. The logic is that if the model is valid, then the actual value of the unknown parameter must be close to the value required to make the model work. If the parameter can subsequently be measured experimentally, this procedure provides a test of the model, and the model has generated a testable hypothesis. If the measured value is not close to the required value, clearly the model would have to be modified or rejected, like any other scientific hypothesis.

Computer realization

The program that we presented earlier in this chapter can provide a useful example for many dynamic state variable models. Here, we briefly note certain kinds of technical issues that will be treated in more detail in the next chapter.

1. *Discretization.* Suppose that the fitness function $F(x,t)$ depends on a single state variable x, where $x_{crit} \leq x \leq x_{max}$, that can take any value in its range (i.e., the state is a continuous variable). For representation in the computer, the values of x must be represented as integers. Choosing the appropriate number of discrete steps n_x between x_{crit} and x_{max} is part of the modeler's task (see Hilborn and Mangel 1997 for some examples)—we usually choose $n_x = 10$, or 20, or up to 100 and run the program using different values of n_x to see how much difference it makes. The relationship between the computer values $(j = 0, 1, \ldots, n_x)$ and the corresponding values of $x = x_j$ is expressed as

$$x_j = x_{crit} + \frac{x_{max} - x_{crit}}{n_x} j \quad (j = 0, 1, \ldots, n_x) \qquad (1.26)$$

Then the dynamic programming equation is solved by the computer for each value of x_j.

2. *Interpolation.* Continuous values of the state variable introduce an additional technicality because the values stored in the computer must be integers. In the next chapter, we describe a simple method of linear interpolation to find fitness for values of the state that are not integers.

3. *Continuous decision variables.* In some models, decision variables may also be continuous. An example is the proportion of a day spent foraging. For the computer, such variables must again be discretized. No interpolation is required here, however. The optimal decisions are determined from the dynamic programming equation for the discretized values of the state variable. On the other hand, how many discrete allocation values to use (10 or 100, for example?) is a matter of concern with the model. We recommend that you begin with a relatively coarse discretization (e.g., running from the minimum to the maximum allocation in 10 steps) and then increase the fineness of the scale once the model is working.

Obtaining predictions from the model

Computer solution of the dynamic programming equation yields the fitness function $F(x,t)$ and the optimal decision function $D(x,t)$. Raw output consists of a printout of these functions, in the form of matrices. But other forms of representing model output are often more useful. Qualitative features of model predictions are immediately transparent from a well-designed graph. Graphs are vastly more efficient in representing data than tables. By arranging several graphs on the same page, parameter sensitivity can also be displayed.

One may wish to know how certain state variables such as fat reserves and body size change over time for an animal that employs optimal decisions. Such state variable trajectories can be computed by **Monte Carlo simulation** or **forward iteration** from the decision matrix, as explained in the next chapter. State variable trajectories (or, in some models, stationary values of state variables) are important predictions of a dynamic state variable model, which can be tested against field data. For example, a model of migration strategy of western sandpipers (Clark and Butler 1999; see chapter 6) generated predictions of the timing of migration of this species from Central America to Alaska, which could be compared with observations.

Summary

Dynamic state variable models are rich in generating testable predictions, which can be compared with available data or used to suggest new experiments. These predictions include behavior and its dependence on internal and external states, as well as its sensitivity to environmental and ecological parameters. Similarly, life-history traits such as growth, timing and level of reproduction, and habitat changes can also be predicted and tested. All these predictions can be displayed graphically in creative ways. In all of this, the computer plays an invaluable role.

1.8 A model for feeding by the three-spined stickleback

Feeding behavior is one of the classic areas of study in behavioral ecology; it is easy to observe and sometimes has clear fitness consequences (starvation), although at other times (for example, animals near satiation) the situation is not that simple. Furthermore, theories of diet selection based on rate-maximizing approaches were remarkably successful (Stephens and Krebs 1986).

On the other hand, it was recognized early on that there was a problem of "common currencies" when organisms risked predation during feeding. Indeed, some of our earliest work (e.g., Clark and Mangel 1986; Mangel 1986; Mangel and Clark 1986) was in part an attempt to resolve the problem of common currencies by using state variable models. Subsequently, Abrahams and Dill (1989) developed methods to measure an energetic equivalence of predation risk. Their work is very elegant and worth reading in the primary literature. Also, see Lima (1998).

Questions of diet choice include (1) which prey items to take, (2) where to feed, and (3) how to respond to predation while foraging. Here we focus on the first question. In the classic theory of diet choice (Stephens and Krebs 1986), prey types $i = 1, \ldots n$ are characterized by net energy gain Y_i, handling time τ_i, and encounter rate λ_i. There are three strong predictions from this theory (which is reviewed in the appendix to this chapter): (1) items are ranked by individual profitability, (2) items are either always included in the diet or always excluded (the 0/1 rule), and (3) encounter rates with excluded items do not affect diet choice.

In the 1970s and early 1980s, Earl Werner and his colleagues, in a series of elegant experiments, set out to test the predictions of rate-maximizing diet selection (e.g., Mittelbach 1981; Werner and Hall 1974; Werner and Mittelbach 1981; Werner et al. 1983a,b; Werner and Hall 1988) using bluegill sunfish *Lepomis macrochirus*. For example, Werner and Hall (1974) used *Daphnia pulex* as prey, for which encounter rates and handling times were estimated by conducting many laboratory feeding experiments. Encounter rates were adjusted in seven different experiments by changing the density of prey items. The range of densities was 20 to 350 individuals per size category.

The results of this experiment agreed qualitatively with the rate-maximizing theory. At low densities of prey items, the fish ate all classes of daphnia in proportion to their abundance. As the encounter rates with the more profitable categories increased, the fish shifted diet to those items, regardless of the encounter rate. However, the 0/1 rule was not upheld: the animals kept some of the less profitable prey items in their diets, in contrast to the theoretical prediction. This raises an important point. The 0/1 rule is a "knife-edge" prediction, and in such a case any variation disproves the theory. But what, exactly, is being disproved? The entire approach (either the use of optimality or the assumption of rate maximization) could be wrong, or the animals could make errors (because they operate with incomplete information). For example, rate maximizing is only a proxy for fitness, and the theory provides no method for evaluating how suboptimal the observed behavior is. One could calculate the rate of energy intake for the observed behavior but would not know if a suboptimal intake rate has meaningful fitness consequences; that is, the fitness consequences of the deviations from the 0/1 rule may be insignificant (Mangel 1989). However, in general, the rate-maximizing model was certainly a better representation of the behavior than a "null" model of random diet choice (also see Werner et al. 1983a for discussion and comparison with other models of diet choice).

Hart and Ison (1991) studied the prey choice of the three-spined stickleback *Gasterosteus aculeatus* L. feeding on *Asellus* (fig. 1.5, table 1.3). The entire story is wonderful and complex; we highlight only part of it and suggest that you visit the literature for the rest (Hart and Ison 1991; Hart and Gill 1992a,b; Hart 1994; and Salvanes and Hart 1998). Two treatments were used to test the rate-maximizing diet choice model, which Hart and Ison called the Basic Prey Model (BPM). The treatments differed in the rate at which prey items were presented to the fish (table 1.4). According to the rate-maximizing theory, items of size 4, 5, and 6 mm should be included in the diet for Treatment A, and items of size 4, 5, 6, and 7 mm should be included in the diet for Treatment B. This was not the case (fig. 1.6). Two things need to be explained. First, why were the smallest prey items almost always accepted by the fish? Second, why were larger prey items sometimes accepted?

When Hart and Ison analyzed the data on acceptance of prey as a function of energy already consumed (fig. 1.7), they discovered a clear pattern. Although the smaller prey items were almost always included, as prey consumption increased, the chance of a larger prey item being accepted decreased. A

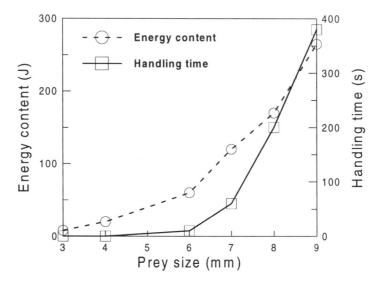

Figure 1.5 Energy content and handling time of *Assellus* prey taken by five stickle-backs (Hart and Ison 1991). A parabola $y = 49.2 - 31.05x + 5.997x^2$ fits the energy content data very well ($r^2 = 0.99$).

Table 1.3 Prey size and energy per unit time per prey for the stickleback–*Assellus* system (Hart and Ison 1991)[a]

Prey size (mm)	Energy content (J)	Energy/time per prey item (J/s)		
		Mean	S.D.	Median
3	7.90	6.91	5.02	4.99
4	15.70	11.96	9.59	9.82
5	58.60	29.54	33.84	18.91
6	80.00	—— Not reported ——		
7	122.40	34.03	53.27	8.80
8	171.70	12.96	23.38	1.81
9	264.50	10.30	14.56	1.96

a. Based on five sticklebacks. Energy/time was calculated as energy content/(pursuit time + handling time). Pursuit time (about 0.7 s) did not vary across prey size.

stepwise multiple regression relating the probability P_i that a prey item was accepted to its size S_i, energy already consumed E, and encounter rate λ_i showed that

$$P_i = 1.52 - 0.16S_i - 0.0009E + 0.11\lambda_i \tag{1.27}$$

Table 1.4 Encounter rates with prey items
for the experiment of Hart and Ison

Prey size (mm)	Encounter rate (min^{-1})	
	Treatment A	Treatment B
3	0.19	0.12
4	0.31	0.28
5	0.44	0.08
6	0.31	0.08
7	0.25	0.09
8	0.36	0.30
9	0.15	0.10

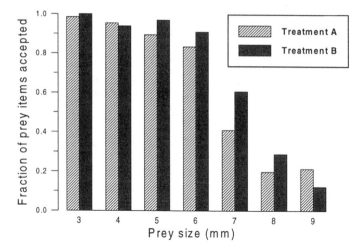

Figure 1.6 The fraction of prey items of different sizes accepted by fish in the experiments of Hart and Ison (1991). For treatment A (striped), the rate-maximizing diet consists of prey categories 4, 5, and 6 mm; for treatment B (grey), the rate-maximizing diet consists of prey categories 4, 5, 6, and 7. Note the problem: small prey items are always accepted, and large ones are sometimes accepted, even though the rate-maximizing theory predicts that they will be rejected.

Hart and Ison (1991, p. 371) concluded that

> The BPM did not predict prey choice under conditions which satisfied most of the model's assumptions. The BPM concentrates on a time-independent prediction of prey choice, ignoring the internal state of the fish.... Our results suggest that it is easier to understand the stickleback's behaviour if we assume that the decision a fish makes over a particular prey is a function only of its present internal state and the size of the prey. The only reference to

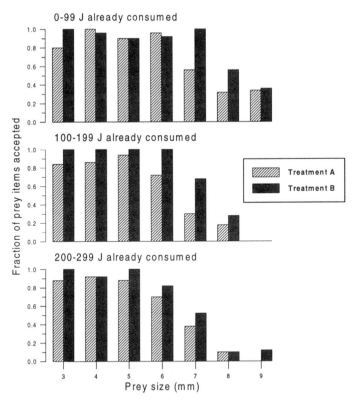

Figure 1.7 The probability that a prey item was accepted depended upon the amount of energy that a fish had already consumed. As before, stripes are treatment A, and greys are treatment B (from Hart and Ison 1991).

future events is through the way in which the current decision affects its future survival.

Hart and Gill (1992a) referred to this as "Werner's paradox": for bluegills the BPM makes predictions of diet choice that were confirmed by the experiment, whereas for sticklebacks the BPM makes predictions that are inconsistent with the experiment. Part of the resolution of the paradox comes from a simple comparison of the predators and their prey (fig. 1.8). A single daphnia is unlikely to change the internal state of a bluegill sunfish. However, a 6–9 mm *Asellus* is indeed likely to affect the gut level of a three-spined stickleback.

This paradox stimulated detailed experimental work (e.g., Hart and Gill 1992b; Gill and Hart 1996) and the development of a dynamic state variable model for the feeding behavior of sticklebacks (Hart and Gill 1992a), which we discuss now.

The state variable in the model is

$$X(t) = \text{stomach contents (in J) at the start of period } t \qquad (1.28)$$

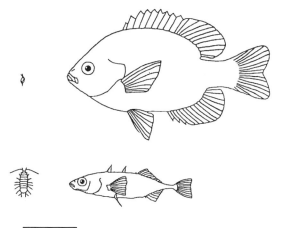

Figure 1.8 A comparison of the predators and their prey (Hart and Gill 1992a). Drawn to scale are a 75 mm bluegill sunfish and its 3 mm Daphnia prey and a 45 mm stickleback and its 8 mm *Asellus* prey. The line represents 20 mm. Thus we expect that consuming a single daphnia will not affect the gut contents of a bluegill sunfish but that consuming a single *Asellus* will affect the state of a stickleback.

and the fitness function $F(x, t)$ represents the maximum probability of survival between t and T, given that $X(t) = x$. Because the state measures stomach capacity, it is constrained to be less than a maximum value $x_{\max} = 450$ J.

If no prey item is encountered or if one is encountered and rejected, the state dynamics are expressed as

$$X(t + 1) = X(t) - a \tag{1.29}$$

whereas if prey item type i is encountered and accepted, the dynamics are given by

$$X(t + \tau_i) = X(t) - a\tau_i + Y_i \tag{1.30}$$

where a is the metabolic cost of a single period and τ_i and Y_i are the handling times and energy gained from prey type i. If p_i is the probability that a prey of type i is encountered, the dynamic programming equation for $F(x, t)$ is

$$F(x, t) = (1 - \sum_i p_i)F(x - a, t + 1)$$
$$+ \sum_i p_i \max\{F(x - a, t + 1), F(x - a\tau_i + Y_i, t + \tau_i)\} \tag{1.31}$$

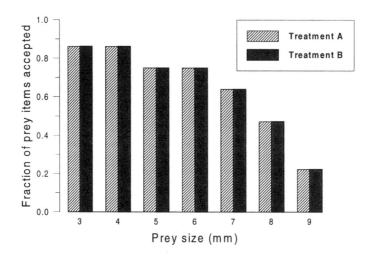

Figure 1.9 A simple dynamic state variable model predicts diet breadth similar to that observed in the experiments of Hart and Ison (1991); compare with fig. 1.6. The results from Hart and Gill (1992a) are based on behavior at $t = 3$, taken across all values of the state variable. Note, however, that the model predicts no difference in diet between treatments A and B (compare fig. 1.6).

with the understanding that the appropriate constraint on the maximum capacity applies. Also, if the state $X(t + 1)$ or $X(t + \tau_i)$ in eqs. 1.29 or 1.30 falls below the critical level $x_{\text{crit}} = 0$, it is assumed that the fish dies of starvation. This can be expressed by writing

$$F(x, t) = 0 \quad \text{if } x < x_{\text{crit}} \tag{1.32}$$

The computer realization of eq. 1.31 includes an IF statement to account for this possibility.

Note that eq. 1.31 is different from the patch choice model. In this case, a prey item of size Y_i is encountered with probability p_i and, given the encounter, the choice is to eat it or not. The term $(1 - \sum_i p_i)$ is the probability that no prey item is encountered in period t, and the summation over i allows us to consider what happens when each prey item is encountered.

Hart and Gill used the solution of eq. 1.31 to mimic the experiments of Hart and Ison (1991) that were summarized in figs. 1.6 and 1.7. The result (fig. 1.9) is moderately encouraging: the model leads us to predict that prey of all sizes will be attacked. However, there is no difference between the two diet treatments (compare figs. 1.6 and 1.9). Thus, there is qualitative agreement in one prediction, but disagreement in another.

Hart and Gill reasoned that eq. 1.31 lacked another key feature of the biology. Larger prey are harder to subdue—a 3 or 4 mm prey can always be captured, but 9 mm prey were captured only 12% of the time (table 1.5). If C_i denotes the probability that an attacked prey is successfully captured, now

Table 1.5 Probability C_i of successfully capturing a prey (from Hart and Gill 1992a)

Prey size (mm)	C_i
3	1.00
5	0.92
6	0.47
8	0.24
9	0.12

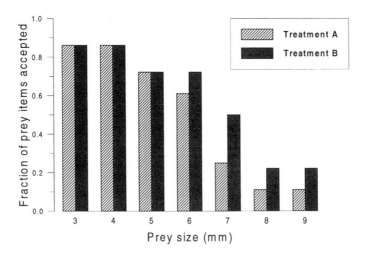

Figure 1.10 A more complicated model (eq. 1.33), which takes the probability of successful attack into account, leads to predictions that are in even better qualitative agreement with the experimental results.

the dynamic programming equation becomes.

$$F(x,t) = (1 - \sum_i p_i)F(x-a, t+1)$$

$$+ \sum_i p_i \max\{F(x-a, t+1), \qquad (1.33)$$

$$C_i F(x - a\tau_i + Y_i, t+\tau_i) + (1 - C_i)F(x-a, t+1)\}$$

Incorporating this feature into the model greatly improves the qualitative fit between the model predictions (fig. 1.10) and the experimental results (fig. 1.6). Furthermore, the state dependence of the pattern (as in fig. 1.7) was

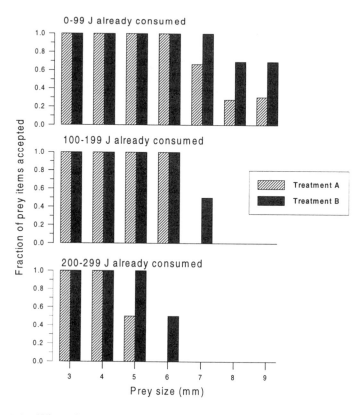

Figure 1.11 When the predictions of the more complicated model are separated according to value of the state variable, the results are qualitatively similar to those observed in the experiments (compare fig. 1.7).

mimicked by the model (fig. 1.11). Hart and Gill also conducted a sensitivity analysis of their predictions, considering how manipulations either expanded or contracted the breadth of the diet, and confirmed the generality of their results.

In summary, the dynamic state variable models agree qualitatively with the experiments. The line of thought suggested by the models continues to lead to new and exciting experiments (e.g., Gill and Hart 1996).

1.9 The role of dynamic state variable models in ecology

The main theme of this book is using dynamic state variable models to improve one's understanding of the relationship between individual behavior and the structure and dynamics of ecological systems. Each of the application chapters, chapters 4–11, describes one or more dynamic state variable models. Many of these chapters also discuss other, simpler models such as rate-maximizing models (see also the appendix to this chapter) and time-minimizing models. These latter models are often elegantly simple; they can provide clear, testable predictions. Then the question is, why bother to develop a fairly complicated dynamic state variable model, if simpler approaches seem to work? As the preceding section 1.8 shows, both types of models have advantages and limitations; this will be further demonstrated in later chapters. The questions of confronting models with data and of evaluating competing models of the same phenomenon are discussed in chapter 3.

To be specific, the following are the main advantages of dynamic state variable models over previous, simpler models:

1. Dynamic state variable models are more flexible than rate-maximizing and similar models. They allow for the consideration of greater biological detail, which may help to explain discrepancies between model predictions and observations. The structure and parameters of a dynamic state variable model are usually tailored to the species and behavior in question.
2. The "currency" assumed to be maximized in a dynamic state variable model, namely (in most cases), expected lifetime reproductive success, is closely related to the basic Darwinian notion of phenotypic fitness. It is the same currency normally used in life-history theory (see chapter 12). Hence, dynamic state variable models provide a close link between two major branches of evolutionary ecology, life-history theory, and behavioral ecology (Clark 1993).
3. Because of 1 and 2 above, dynamic state variable models are unrestricted in principle in their ability to handle multiple trade-offs, variable environments, anticipatory and informational aspects of behavior, and other details. By working with a sequence of progressively more complex models, you can assess the importance of different components of the organism's phenotype and environment.

It may seem surprising that a model based on maximizing ultimate lifetime reproduction could predict microbehavior such as daily foraging routines, oviposition strategy, or migration schedules. Indeed, when we started using dynamic state variable models in the late 1980s, at first we were astounded at their success. Now we have come to expect it. Part of the reason for this success is the flexibility of these models (see chapter 2). However, we do not agree with the criticism that these models are too flexible, capable of predicting anything—we have experienced enough failures to know that merely fiddling with the parameters of an inappropriate model is not likely to work. Dynamic state variable models are eminently testable, as later chapters will demonstrate.

We stress that one should not be wedded to technique: dynamic state variable models are not always superior to rate-maximizing or any other type of model. Simplicity and generality are the main advantages of rate-maximizing models and the like. Such models can often add much to one's understanding of the effects of ecological parameters on individual behavior.

In many cases the development and analysis of such simple models may be the first step in the theoretical component of a research program, and more complex models are brought in at a later stage.

Dynamic state variable models are limited in their ability to incorporate certain ecological factors, such as population or community-level effects. Nevertheless, later chapters will indicate that such processes can be included to some extent, usually resulting in a major increase in model complexity. An alternative modeling approach, genetic algorithms, is discussed briefly in chapter 10.

Ecological and behavioral systems are so complex that no single modeling approach is likely to be ideal for all questions. The ability to work with many different kinds of models is a great asset to any biologist interested in adaptation and evolution.

1.10 Examples of projects

We have found that students learn a lot at this stage by developing an original dynamic state variable model. It is particularly useful to do this in groups, including the whole class. The following are examples of projects developed by students in the courses that we or colleagues have taught in recent years. A model based on one of these descriptions (or an original one) should be constructed. You will find that the process of reaching agreement on the dynamic programming equation determines whether your model is complete and unambiguous. Programming and using the model should probably wait until you have read chapters 2 and 3.

Adaptation of barnacle life-history traits to mortality due to oil spills
(Samantha Forde, University of California, Santa Cruz).

The rocky intertidal region near Santa Barbara, California is particularly vulnerable to oil spills because of offshore oil drilling and a high amount of tanker traffic. In 1969 this area suffered one of the largest oil spills in California history. Further, the area is characterized by numerous natural subtidal oil seeps, which have been documented as far back as 1792. The oil from these seeps frequently washes ashore and the rocky intertidal in this region often has large amount of oil (tar) on it.

The barnacle *Chthamalus fissus* dalli characterizes the mid and upper zones of the intertidal. The life history and population dynamics of this organism are well-known, and it therefore provides an ideal system to test the implications of oil spills on life history traits of an intertidal organism. A model can be used to evaluate the consequences of a major oil spill assuming that reproductive behavior has evolved under chronic, low levels of tar cover from subtidal oil seeps.

Life history traits are constrained by the trade-off between current reproductive growth, probability of mortality due to tar cover, and future reproductive output. A dynamic state variable model can be used to evaluate the allocation of resources to growth versus reproduction as the probability of mortality increases. The results of the model should provide a threshold state of reproduction (i.e., where risk of mortality is high and resources are no longer allocated to growth) that varies as a function of probability of mortality.

Expected reproductive output can be calculated using forward iterations of the model for low probabilities of mortality (no oil spill) and under high probabilities of mortality (oil spill). The loss of potential larval contribution to the population is the difference between expected reproductive output under these conditions. This difference gives an indication of whether or not the evolution of life history traits under chronic, low levels of tar cover has implications for population dynamics in the event of a major oil spill.

Movement strategies of American marten during the winter
(Kim Heinemeyer, University of California Santa Cruz)

American marten (*Martes americana*) are small to mid-sized carnivores in the weasel (Mustelidae) family. They are habitat specialists of temperate and boreal coniferous forests, primarily using mature and older age classes of trees, and avoiding open areas. Within these forests, marten prey upon small mammals such as mice, voles, and tree squirrels. Marten remain active during the winter, during which they are under extreme energetic stress. Snow and cold temperatures dramatically increase the cost of travel and thermoregulation, and limit availability of prey. Additionally, marten are physically and physiologically poorly adapted to deal with winter extremes: their long, thin, weasel-like body form permits high heat loss, their short fur provides limited insulation, and they do not store energy in the form of either fat or food caches. They survive winter by traveling across the surface of the snow (the coldest microclimate in the boreal forest), searching for access sites that provide entrance to the under-snow (subnivean) air spaces, where they find both prey in the form of small mammals and thermally protected resting sites. Marten maintain territories through the winter months, and this places additional demands upon their limited energy resources. Territory maintenance is probably directly linked to future reproductive success, particularly for females, who maintain territories that will provide sufficient resources for kit raising.

Dynamic state variable models may be used to examine the trade-offs marten face during the critical winter months. One trade-off is between energetically costly territory monitoring, which will ensure future reproductive success, and immediate energetic needs of winter survival. Movement behaviors occur on a daily basis. By not moving, a marten avoids the high cost of movement, but gains little in energy, because of the scarce availability of prey at the already occupied site. Alternatively, movement may be limited to simply finding another access site to a new supply of prey, minimizing travel costs but not providing any benefits of territory monitoring. Finally, a marten may travel long distances for the purposes of monitoring and defending its home range. This option entails high energetic costs, but will have significant benefits to future reproductive success and also allows the marten to find a new access site and supply of prey. The cumulative survival and fitness probabilities through the winter will determine the marten's fate and reproductive fitness. These can be optimized in a state variable model, thus providing insights into marten winter strategies.

It is relatively simple to add complexity to the model. One option is to examine the change in marten winter strategies with fragmentation of their forest habitat. Timber harvest, particularly clear cutting, has pronounced impacts on martens by directly removing habitat and also by fragmenting the remaining habitat. Martens avoid clear-cut areas and generally travel around them. This significantly increases the distances moved by martens as they travel through their home ranges for food and territorial monitoring. This extra travel translates into increased energetic stress on the animal, which may be actualized in the decision to cross clear-cut areas at the high risk of predation rather than travel around the area. Or potentially, the marten may choose not to monitor all of its territory because this would entail excessively long movements as it skirts clear-cut areas. Then this would have a strong influence on its future reproductive success. The results of a modeling effort may provide insights into the impacts of habitat fragmentation due to forest management practices and possibly could be applied to land management decisions.

Allocation decisions for plants (Tim Kailing, Princeton University)

Plants trade off between investing in root biomass or shoot biomass: roots are necessary to acquire nutrients and water, but they are a net energy sink, whereas the photosynthetically active shoot is the energy source for the plant. In Mediterranean annuals, this trade-off is particularly evident: during the distinct rainy season, the soil is saturated, and with just a minimal investment in roots, plants that have greater shoot allocation can grow faster. However, after the rains end, the water recedes, and only those plants that invest heavily in roots—at the expense of current shoot growth—can access the receding water and continue to grow. So, plants that invest in roots are trading off current growth for a longer growing season. What is the optimal allocation strategy for a plant in such an environment? Does the optimal strategy change as the rate of water recession increases? How does the addition of a stochastic shoot herbivory term change the results of the model?

Anuran breeding behavior (Shannon McCauley, University of Florida)

Frogs often form dense aggregations during their breeding seasons. In most of these species, males use vocalizations to attract females, who may be able to make some assessment of male quality on the basis of these calls. The call parameters females assess differ among species but include the loudness of the call, call duration, number of calls produced in a night, and call frequency. Some of these parameters are largely a function of male size (e.g., large frogs typically produce the lower frequency calls favored by females), whereas others may depend on the energy an individual has to invest in calling over a given night (e.g., number of calls produced and call duration). The energetic costs of producing calls can be high, and calling attracts predators, as well as females; consequently calling males face higher predation risks. Some males attempt to avoid these costs by being satellite males. They sit near the calling males and attempt to intercept approaching females. They have lower probabilities of acquiring mates than calling males, but their energetic costs and predation risks are also reduced.

A model can be used to assess the trade-off across strategies (calling, being a satellite, foraging, and hiding) for males during the breeding season and to examine factors that affect their breeding phenologies, where the fitness criterion is to maximize the sum of all breedings for one season plus a small terminal fitness. Once the model is constructed, one can determine the optimal behavior for each state. Following this behavior forward in time allows determining the proportion of males in the population that adopt calling and satellite strategies at each time, which then is testable in the field.

Parasitic plants (David Smith, Princeton University)

Roughly 1% of flowering plants are parasitic and receive some portion of their resources by penetrating the vascular tissue of the root or stem of another plant. For root parasitic plants, the benefits of attaching to a host plant include huge increases in seed production or other proxies of fitness. For this reason, allocation decisions in root parasitic plants may differ from allocation decisions in ordinary plants.

To understand these differences, construct a model of the life history of *Orthocarpus tolmei*, a small, annual, root-parasitic plant. The model has four state variables: biomass of roots, shoots, and corolla, and the number of hosts to which the plant has attached. Each day, the plant has metabolic costs that are proportional to total biomass. Assume that energy production (expressed in the same units as biomass) is limited by a single resource and that the total amount of resources the plant receives in a day is the sum of two terms. The first term is proportional to the amount of root biomass, and the second term is proportional to the number of hosts. The gross energy production is a function of total resources acquired and total shoot biomass. At the end of each day, the net energy production is allocated to root, shoot, or corolla. The parasitic plant increases its host status by one if it encounters a host and produces new haustoria. The likelihood of encountering a new host is a function of standing root biomass (if a host grows into the parasitic plant's roots), as well as allocation to new roots (if it grows into a space occupied by a host plant's root). Finally, the fitness of a parasitic plant is its total allocation to corolla. The length of the growing season is stochastic, beginning around July 9 and ending around September 1.

Once the model has been constructed, use it to answer the following questions: (1) How does allocation differ depending on whether the potential for parasitism is present? (2) How does the total biomass of a plant change if it attaches early in the growing season compared to late in the growing season? (3) What is the expected effect on total biomass of killing the host plant, either early or late in the growing season?

Nest departure in Alcids (Ron Ydenberg, Simon Fraser University)

(Note: this is more a problem set than an original project, but you may enjoy working on it.)

The seabird family Alcidae (the murres, puffins, auklets, etc.) is unique among birds in that species exhibit widely varied developmental modes. There are three main patterns. Most species (15 of 22) are "semiprecocial" and grow to about 80% or more of final adult body mass on food provisioned to the nest by their parents. Three "intermediate" species grow to about 25% of adult body mass in the nest, whereupon they depart to complete their development at sea. Many of these species have been well studied (e.g., Nettleship and Birkhead 1986). In the four remaining species, chicks are precocial, depart the nest as 1- or 2-day-old downy chicks, and both parents tend them at sea (Gaston 1992).

Ydenberg and his colleagues have used dynamic state variable models to examine how natural selection has shaped this diversity (Ydenberg 1989; Clark and Ydenberg 1990a,b; Ydenberg et al. 1995). These models consider nest departure in the semiprecocial and intermediate species using the following framework: a juvenile's life is considered to be composed of two phases, nest and ocean, that are successively occupied as it grows. Both phases have associated growth and mortality characteristics, and the models calculate the nest departure (i.e., age and mass at transition from nest to ocean phase) policy that maximizes fitness.

In this problem you're asked to write dynamic programming equations for a variety of extensions to the basic model presented in Ydenberg (1989). You should read this paper and understand the dynamic programming equations (eqs. 1–7) before doing this problem. (There are some minor notational errors in the equations.) The paper considers one of the intermediate species, the common murre *Uria aalge*. Both murre parents participate in rearing a single chick to 15–25 d. Then the chick departs, and jumps or glides from the ledge into the sea, where it is accompanied by its father for a further 2 months or so while it completes development. At nest departure, it has attained about one-quarter of adult body mass and has replaced its down with body plumage, but it has no wing or tail feathers. There is extensive intraspecific variation in the age and mass at departure, discussed further by Ydenberg et al. (1995).

The basic model covers a 90-d breeding period, and the state variable is body mass x. The chick is assumed to grow according to a deterministic function $g_n(x)$ while in the nest, and according to a second function $g_o(x)$ while at sea. The daily mortality while in the nest is m_n, and while at sea is m_o. The chick is assumed to undergo a mortality risk f in making the transition. For reasonable parameter values estimated from the literature, the nest departure age and mass predicted by the model agreed well with what is observed (but see Gaston 1998). As Byrd et al. (1991) point out, when the problem is formulated completely deterministically as it is here, there is no need for a dynamic state variable model. The fitness of departing on each of the 90 days of the breeding season can easily be computed, and the optimum found. But a dynamic state variable approach becomes essential as soon as any variability is introduced. To appreciate this point, modify Ydenberg's dynamic programming equation to accommodate each of the following situations:

1. Suppose that there are good and bad years that influence the ocean rate of growth. Assume that the nest rate of growth is not influenced by the state of the ocean and that in deciding when to depart, the nestling knows what type of year it is. (This could be communicated by the parents.)
2. Next suppose that the nestling does *not* know what type of year it is but does know the frequency with which good and bad years occur.

A further modification would be to suppose that the nestling begins its life with a prior expectation based on the frequency of good and bad years but updates this expectation each day based on information received from its parents, so that its estimate of whether the year is good or bad is based on more information as it ages. The methods of Bayesian information updating described in chapter 11 can be used here, and you should study that chapter if you wish to consider this possibility. Specifically, suppose as before that there are good and bad years at sea, but now assume that, in addition to affecting the chick's growth after nest departure, this variation also influences the amount of food that the parents deliver. The variable delivery not only affects the nestling's estimate of the type of year but also its growth. Modify Ydenberg's model to see how much all the extra complexity alters the original solution.

Appendix: Rate-maximizing theories of diet selection and habitat choice

Diet selection

In the classic theory of diet choice (Stephens and Krebs 1986), prey types $i = 1, \ldots n$ are characterized by net energy gain Y_i, handling time τ_i, and encounter rate λ_i. The profitability of an individual prey item is measured by the rate of flow of energy to the organism, which is Y_i/τ_i.

We now derive the rate-maximizing result in detail for the case of two prey items. In this case, the more profitable prey item is always eaten, and the question is when the less profitable prey should item be eaten. First, suppose that the organism specializes: it ignores a prey item of type 2 when encountered, because it does not want to pay the opportunity cost of lost search time spent eating the prey item of type 2. Imagine a long time interval T that consists only of search behavior and handling behavior. If these take time S and H, respectively, then,

$$S + H = T \qquad (A.1)$$

Since the encounter rate with prey items of type 1 is λ_1 items per unit time, the average number of prey items encountered in search time S is $\lambda_1 S$. The total handling time associated with these prey items is $H = \tau_1 \lambda_1 S$, so that eq. A.1 becomes

$$S + \tau_1 \lambda_1 S = T \qquad (A.2)$$

from which we conclude that

$$S = \frac{T}{1 + \tau_1 \lambda_1} \qquad (A.3)$$

The total energy obtained from these prey items is the number of prey items multiplied by the energy per prey item; thus, it is $\lambda_1 S Y_1$. Using eq. A.3, the total energy obtained from prey captures over the interval of length T, if the organism specializes, is given by

$$E_1 = \frac{\lambda_1 Y_1 T}{1 + \tau_1 \lambda_1} \qquad (A.4)$$

so that the rate of energy gain if the organism specializes, is expressed by

$$R_1 = \frac{E_1}{T} = \frac{\lambda_1 Y_1}{1 + \tau_1 \lambda_1} \qquad (A.5)$$

If the organism generalizes, it accepts either prey item upon encounter. You should follow the analysis leading to eq. A.5, modified for two prey items, to show that the rate of energy gain when both prey items are accepted is given by

$$R_{1,2} = \frac{\lambda_1 Y_1 + \lambda_2 Y_2}{1 + \tau_1 \lambda_1 + \tau_2 \lambda_2} \qquad (A.6)$$

Thus both prey items should be included in the diet if $R_{1,2} > R_1$ and vice versa.

In general, if k prey items are included in the diet, the rate of return $R_{1,\dots,k}$ is given by

$$R_{1,\dots,k} = \frac{\sum_{i=1}^{k} \lambda_i Y_i}{1 + \sum_{i=1}^{k} \tau_i \lambda_i} \qquad (A.7)$$

and the optimal diet consists of including items $1, \ldots, i*$, found by computing the rates in eq. A.7 until the optimal energy return is found. This predicts the optimal diet.

There are three strong predictions from this theory: (1) items are ranked by individual profitability, (2) items are either always included in the diet or always excluded (the 0/1 rule), and (3) encounter rates with excluded items do not affect diet choice.

Habitat choice

The rate-maximizing theory of diet selection is based on the idea that, while foraging, an organism is expected to maximize its average rate of energy intake—see eqs. A.5–A.7. Whether this is the best fitness "currency" to use in studying foraging behavior has been widely discussed; see the book of Stephens and Krebs (1986) and the article of Ydenberg et al. (1994) for an introduction to this literature. Here we raise the question of a fitness currency for the situation where a forager can choose between n foraging habitats that yield different rates of food intake Y_i and impose different rates of predation risk m_i. We will show, by an easy argument, that under simplifying assumptions, the optimal habitat is the one that minimizes the ratio R_i of risk to benefit:

$$R_i = \frac{m_i}{Y_i} \qquad (A.8)$$

The earliest derivation of a rule of this kind is due to Werner and Gilliam (1984), who modeled ontogenetic habitat shifts.

Consider a forager that makes repeated foraging trips to foraging habitats i that have different rewards and predation risk:

$$m_i = \text{risk of predation per trip}$$
$$Y_i = \text{average food recovered per trip}$$

Assume that the number of possible trips is unlimited, so that the optimal habitat will be the same for all trips. If habitat i is used, total lifetime expected food recovery is given by

$$Y_{\text{total}} = (1 - m_i)Y_i + (1 - m_i)^2 Y_i + \cdots = \frac{1 - m_i}{m_i} Y_i \qquad (A.9)$$

(from the formula $1 + x + x^2 + \cdots = \frac{1}{1-x}$ for the sum of a geometric series).

Now make the additional assumption that all $m_i \ll 1$. Then, $1 - m_i \approx 1$, and eq. A.9 becomes

$$Y_{\text{total}} \cong \frac{Y_i}{m_i} \qquad (A.10)$$

The optimal habitat i maximizes this expression, or, what amounts to the same thing, it minimizes

$$\frac{m_i}{Y_i} \qquad (A.11)$$

Several simplifying assumptions underlie this derivation. First, there are no time constraints (such as the end of summer). Second, foragers do not undergo senescence. Third, fitness is assumed proportional to the total food obtained. Finally, to obtain the approximation in eq. A.10, it is assumed that predation risk per trip is small.

2

Some Details of Technique

In this chapter we discuss certain important techniques relevant to dynamic state variable models.

2.1 Linear interpolation

The patch-selection model of chapter 1 treated the state variable x in terms of discrete values, $x = 0, 1, 2, \ldots, x_{\max}$. Changes in x due to food intake and metabolic costs were also given in the same discrete units. In the real world, state variables often vary continuously over some range such as $0 \leq x \leq x_{\max}$. For example, we might measure the size of a fish in mm, with a maximum size of 200 mm. Growth rates $g(x)$ are known, in mm/day. Then we might naively allocate 200 locations in memory for the size variable. However, there are two reasons why this is not the best approach.

First, the daily growth $g(x)$ will not necessarily be a whole number of mm—how would we represent $X(t + 1) = x + g(x)$ if $g(x) = 0.37$ mm? We cannot simply round off to the nearest mm, as this would be paramount to assuming zero growth rate in this case. Rounding upward to $x + 1$ would be equally inaccurate. Clearly we need to **interpolate** between x and $x + 1$; fortunately this is easy to do, as we explain here.

An alternative to interpolation might be to use smaller discrete units for x, for example, 0.01 mm. This would require 20,000 storage locations in the computer (and 20,000 computations of $F(x, t)$ for each time step t). This is very inefficient. Worse, in a model with several state variables, the method would simply not be feasible. The method of linear interpolation resolves these difficulties.

Suppose that the fitness function $F(x, t)$ depends on a single state variable x, with

$$x_{min} \leq x \leq x_{max} \tag{2.1}$$

where x is a continuous variable, capable of taking on any value in this range. For the computer, x must be discretized and the values of $F(x, t)$ represented in the computer by a finite array, say Fit$[j, t]$, where

$$j = 0, 1, \ldots, n$$
$$t = 1, 2, \ldots, T \tag{2.2}$$

The equation

$$x_c = \frac{n}{x_{max} - x_{min}} (x - x_{min}) \tag{2.3}$$

converts the original ("real-world") value of x into the computer value x_c (which is not necessarily an integer value). Conversely, the equation

$$x = x_{min} + \frac{x_{max} - x_{min}}{n} x_c \tag{2.4}$$

(obtained by solving eq. 2.3 for x) converts x_c back to x. Note, in both equations, that $x_c = 0$ corresponds to $x = x_{min}$ whereas $x_c = n$ corresponds to $x = x_{max}$. These equations can be used in your computer code to convert back and forth between the values of x and x_c.

Now we write

$$\text{Fit}(x_c, t) = F(x, t) \tag{2.5}$$

noting that only the values Fit(j, t) $(j = 0, 1, \ldots, n)$ are actually stored in the computer. To obtain the values of Fit(x_c, t) when x_c is not an integer j, we use linear interpolation. Specifically, if

$$j \leq x_c < j + 1 \tag{2.6}$$

we write

$$\Delta x_c = x_c - j \tag{2.7}$$

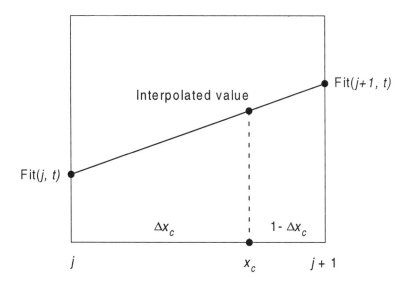

Figure 2.1 Linear interpolation in computer coordinates (see text).

Then (see fig. 2.1),

$$
F(x,t) = \text{Fit}(x_c, t) \cong \begin{cases} (1 - \Delta x_c)\text{Fit}(j,t) + \Delta x_c \text{Fit}(j+1,t) & \text{if } j < n \\ \text{Fit}(n,t) & \text{if } j = n \end{cases}
$$

$$(2.8)$$

The sample codes available on the Web site contain examples of linear interpolation.

How many discrete values (n) of x_c should be used? Typically we choose $n = 20$, 50, or 100, using smaller values of n while developing the code, and then a larger value for the final computations. (Of course the computer code should be written with n as a parameter, which can then be set to the desired value.) Linear interpolation with $n = 100$ provides sufficient accuracy for most cases; pure discretization without interpolation would require at least $n = 10,000$ for comparable accuracy.

Two-dimensional interpolation

When there are two continuous state variables x and y, with

$$
x_{\min} \le x \le x_{\max}, \quad y_{\min} \le y \le y_{\max}
$$

then y is also converted to the computer value y_c:

$$
y_c = \frac{m}{y_{\max} - y_{\min}}(y - y_{\min})
$$

and conversely

$$y = y_{\min} + \frac{y_{\max} - y_{\min}}{m} y_c$$

We write

$$\text{Fit}(x_c, y_c, t) = F(x, y, t)$$

and note that the values $\text{Fit}(j, k, t)$ $(j = 0, 1, \ldots, n;\ k = 0, 1, \ldots, m)$ are actually stored in the computer. If

$$k \leq y_c < k + 1$$

we write

$$\Delta y_c = y_c - k$$

The linear interpolation formulas are

$$
\begin{aligned}
F(x, y, t) &= \text{Fit}(x_c, y_c, t) \\
&\cong (1 - \Delta x_c)(1 - \Delta y_c)\text{Fit}(j, k, t) + (1 - \Delta x_c)\Delta y_c \text{Fit}(j, k + 1, t) \\
&\quad + \Delta x_c(1 - \Delta y_c)\text{Fit}(j + 1, k, t) + \Delta x_c \Delta y_c \text{Fit}(j + 1, k + 1, t) \\
&\qquad \text{if } j < n \text{ and } k < m
\end{aligned}
\tag{2.9}
$$

and

$$
\begin{aligned}
F(x, y, t) &= \text{Fit}(x_c, y_c, t) \\
&\cong
\begin{cases}
(1 - \Delta y_c)\text{Fit}(n, k, t) + \Delta y_c \text{Fit}(n, k + 1, t) & \text{if } j = n \text{ and } k < m \\
(1 - \Delta x_c)\text{Fit}(j, m, t) + \Delta x_c \text{Fit}(j + 1, m, t) & \text{if } j < n \text{ and } k = m \\
\text{Fit}(n, m, t) & \text{if } j = n \text{ and } k = m
\end{cases}
\end{aligned}
\tag{2.10}
$$

There are analogous (but more complicated) formulas for more than two state variables.

2.2 Sequential coupling

Most organisms experience a series of life-history stages. They also face temporal changes in their environment, including day–night and annual seasonal alternations. Dynamic state variable models are ideally suited for studying adaptations to these temporal changes. Consider, for example, the case of day–night alternation for a diurnal animal. We will modify the basic patch-selection model of chapter 1 to cover this situation.

Now we assume that $t = 1, 2, \ldots, T$ denotes the time of day, discretized into T periods of equal duration, and encompassing daylight hours. (We ignore seasonal variation in day length, but this could easily be included.)

Let $d = 1, 2, \ldots, D$ denote day number over the breeding season. Overnight metabolic cost is c_{night}, a random variable reflecting the possibility of mild versus cold nights.

Next the fitness function is defined as

$F(x, t, d)$ = maximum expected reproduction from period t on day d to the end of the organism's life, given $X(t, d) = x$

$$(2.11)$$

This leads to the terminal condition

$$F(x, T, D) = \Phi(x) \qquad (2.12)$$

At the end of each day $d < D$, the terminal fitness equals the expected fitness starting the next day, accounting for overnight costs:

$$F(x, T, d) = E\{F(x - c_{night}, 1, d + 1)\} \qquad (2.13)$$

where E denotes expectation with respect to c_{night}. (Make sure that you see how this is derived from the basic definition in eq. 2.11.) Equation 2.13 provides the sequential coupling between days and nights.

Finally, the dynamic programming equation for daytime decisions is identical to the patch-selection model (see eq. 1.16)

$$F(x, t, d) = \max_i V_i(x, t, d) \qquad (2.14)$$

where the expressions $V_i(x, t, d)$ are exactly as in eqs. 1.14–1.15. As an exercise you may wish to modify the code for the patch-selection model to include this extension. What effects should the change have on the model's predictions?

Seasonal alternation can be modeled similarly, although in this case behavior in each season may be included in the model. Continuing with the patch-selection model, now let $F_s(x, t, d)$ and $F_w(x, t, d)$ denote summer and winter fitness, respectively. Then

$$F_s(x, T_s, D_s) = E\{F_w(x - c_{night}, 1, 1)\} \qquad (2.15)$$

where T_s and D_s denote the number of periods in a summer day and the number of days in summer. Similarly,

$$F_w(x, T_w, D_w) = E\{F_s(x - c_{night}, 1, 1)\} \qquad (2.16)$$

These two equations couple winter and summer alternatively. The dynamic programming equation for F_s is the same as before. The equation for F_w omits the possibility of reproduction. Also the model parameters (food abundance, predation risk, and metabolic costs) would presumably differ between summer and winter.

One final detail needs to be clarified—where to initiate the backward iterations? Suppose that the animal can live for a maximum of Y years. Then, for the Yth year, eq. 2.15 is replaced by

$$F_s(x, T_s, D_s) = 0 \tag{2.17}$$

and the model is iterated for years $Y, Y - 1, \ldots$, alternating winters and summers. Notice that this grand, full-life model actually has four different time scales: day, night, season, and year. Total lifetime reproduction—that is, fitness—is the sum of reproduction over all future periods. The dynamic state variable framework automatically accounts for total lifetime reproduction and survival.

This raises the question as to whether total lifetime reproduction is always the appropriate measure of fitness for iteroparous organisms. We discuss this topic in chapter 12.

2.3 Forward iteration and Monte Carlo simulation

The raw output from the dynamic programming calculation consists of two arrays, the fitness function $F(x, t)$ and the decision function $D(x, t)$. These can be thought of as quantitative predictions of the model, but testing these predictions may be formidable. For example, assessing the current state of an animal in the wild can be difficult, and measuring fitness is also problematical (Endler 1986).

Fortunately, additional predictions, which may be much more easily tested, can readily be obtained by the method of **forward iteration**. We consider an individual with given initial state $x_1 = X(1)$. Then we compute in turn $X(2), X(3), \ldots$, assuming that the individual always uses the optimal decision $D(x, t)$. These values $X(t)$ are computed assuming the same state dynamics as in the dynamic state variable model.

In most models, however, the state dynamics are stochastic. In the patch-selection model of section 1.4, for example, $X(2)$ will equal zero if the individual is killed by a predator, or will equal $X(1) + Y_i - a_i$ if it is not killed and finds food Y_i, and so on. Suppose that $X(1) = 10$, so that the optimal patch is $i = 2$ (see fig. 1.2), with parameter values $m_2 = .05$, $p_2 = .5$, $Y_2 = 4$, and $a_2 = 1$. It follows that

$$X(2) = \begin{cases} 0 & \text{with probability } .05 \\ 13 & \text{with probability } .475 \\ 9 & \text{with probability } .475 \end{cases}$$

We can interpret this distribution for $X(2)$ in either of two ways. From the individual's perspective, it will die of predation in period 1 with a 5% chance, will increase its reserves from 10 to 13 with a 47.5% chance, etc. Alternatively, in a large population of individuals, all using the optimal strategy and starting out with $X(1) = 10$, 5% will be killed immediately, 47.5% will increase $X(2)$ to 13, and 47.5% will have $X(2) = 9$. To express this mathematically, let

$$P(x, t) = \Pr\{X(t) = x\}$$

for an organism that always uses the optimal decisions $D(x, t)$ and has $X(1) = 10$. Then

$$P(0, 2) = .050$$
$$P(9, 2) = .475$$
$$P(13, 2) = .475$$

and $P(x, 2) = 0$ for $x \neq 0$, 9, or 13. Now the calculation can be continued to find $P(x, 3)$, then $P(x, 4)$, and so on. The procedure is best organized in terms of matrix calculations, and this is known as the method of **Markov chains**: see section 10.2. Here we will describe an alternative approach, called **Monte Carlo simulation**. Either method can be used in practice; most people find Monte Carlo simulation easier to understand. The Monte Carlo method provides approximate values of $P(x, t)$, whereas the Markov chain iteration computes these values exactly. On the other hand, the Monte Carlo method can be used to obtain predictions in addition to the state distribution as obtained from the Markov chain calculation; an example is given here.

Monte Carlo forward iteration

The random number generator included in a given computer language package produces a random (actually pseudorandom) number r each time you use it. These random numbers are uniformly distributed over a certain interval $0 \leq r \leq r_{\max}$; in many compilers $r_{\max} = 1$, which we will assume here. (If not, it is only necessary to divide each r by $r_{\max}$.)

Now suppose that in our dynamic model a certain event (e.g., predation) occurs with probability p. To simulate this, we generate a random number r and assume that:

if $r \leq p$, the event occurs
if $r > p$, the event does not occur

To see how this works, consider the following pseudocode for Monte Carlo forward iteration in the patch-selection model of section 1.4.

1. Initialize $t = 1$, $x = X(1)$ [given], and TotalRep = 0 (TotalRep is total reproduction until now).
2. Find the optimal decision $D(x, t) = i$; let a_i, m_i, p_i, and Y_i be the corresponding patch parameters.
3. (Foraging) If $i = 1$ or 2, choose random number r_1. If $r_1 \leq m_i$ (organism is killed), set $x = 0$, and end simulation. Otherwise, continue.

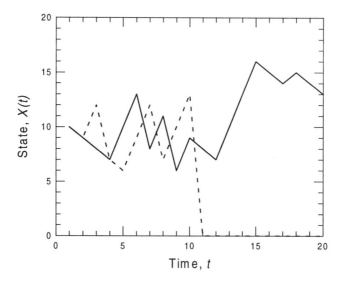

Figure 2.2 Two simulated state trajectories $X(t)$ for the patch-selection model. Solid line: organism survives 20 days and obtains total reproduction of 8 units; dashed line: organism dies of predation on day 11, having achieved total reproduction of 12 units.

4. Choose random number r_2. If $r_2 < p_i$ (organism finds food), set $x = x + Y_i - a_i$. Otherwise, set $x = x - a_i$ (subject to $0 \leq x \leq x_{\max}$). If $x = 0$ (organism starves), end simulation. If $t = t_{\max}$, end simulation. Otherwise, set $t = t + 1$, and go to step 2.

5. (Reproduction) If $i = 3$, then if $x < x_{\mathrm{rep}}$, set fitness_increment $= 0$ and $x = x - a_3$. Otherwise, if $x < x_{\mathrm{rep}} + c$, set fitness_increment $x - x_{\mathrm{rep}}$, and set $x = x_{\mathrm{rep}} - a_3$. Otherwise, set fitness_increment $= c$, and then set $x = x - a_3 - c$ (subject to $0 \leq x \leq x_{\max}$). Set Total Rep $=$ TotalRep $+$ fitness_increment. If $x = 0$, end simulation. Otherwise, continue.

6. Choose random number r_3. If $r_3 < m_3$ (organism is killed), set $x = 0$, and end simulation. If $t = t_{\max}$, end simulation. Otherwise, set $t = t + 1$ and go to step 2.

Thus, during the simulation you keep track of the state $X(t)$ and TotalRep(t). You should also record information needed for later statistical description of the results, e.g., mean, standard deviation, etc. The code Patchnew.cpp (written in C^{++}) available on the Web site includes Monte Carlo simulation for the patch model, following the previous pseudocode. Some sample state trajectories are illustrated in fig. 2.2. How to use these methods in empirical models is discussed further in chapter 3.

2.4 The critical level of reserves

In the patch-selection model of chapter 1, we assumed that the forager would die of starvation if its reserves $X(t)$ fell to a critical level, defined as $x_{crit} = 0$ for simplicity. For actual organisms this assumption may be oversimplified in at least two ways. First, the individual may weaken without necessarily immediately dying at low reserve levels. This possibility could be modeled by including a reserve-dependent survival probability $S(x)$, where $S(x_{crit}) = 0$ and $S(x)$ increases for $x > x_{crit}$. Is it worthwhile to include this level of detail? Perhaps not, unless one is specifically studying adaptations to temporary nutrient shortages, as with desert plants or animals, for example.

Second, when primary reserves such as fats and lipids are depleted, an organism may switch to using secondary tissues such as muscle proteins. The cost of replacing these tissues may be considerably greater than that of replacing primary reserves, implying that the secondary reserves are used only for emergencies. For example, research on diving birds and mammals has shown that under normal conditions, respiration during dives is entirely aerobic. Divers are capable of anaerobic metabolism, however, and occasionally use anaerobiosis on extended dives. Ydenberg and Clark (1989) studied the diving behavior of breeding western grebes (*Aechmophorus occidentalis*), observations of which had clearly indicated anaerobiosis. The main state variable in Ydenberg and Clark's model was $X(t) =$ oxygen or equivalent "debt" (mL of O_2). The range $0 \leq X(t) \leq x_1$ corresponds to actual O_2 depletion in blood and tissues, and the range $x_1 < X(t) \leq x_2$ corresponds to anaerobic conditions. A bird with $X(t) = x_2$ cannot continue diving and must surface. In this model, there are thus two critical levels of reserves, x_1 and x_2. This aspect was the main feature of the grebe model, which demonstrated that (contrary to accepted wisdom in the field) deliberate anaerobic diving could well be optimal under certain natural conditions.

When comparing models to experimental or field observation, the critical value of the state variable may be greater than zero. For example, in models in which the state variable represents body fat, we expect the critical value to exceed zero since not all fat may be accessible as an energetic reserve.

2.5 Environmental state variables

Until now we have assumed that $X(t)$ is an internal (phenotypic) state of the individual. Other state variables are sometimes important determinants of behavior, however. Two important examples are environmental and informational states. Here we discuss the modeling of fluctuating environments; informational state variables are discussed in chapter 11.

In modeling the effect of the environment on behavior, three cases can be identified: (a) the environment is deterministic, (b) it is stochastic without serial correlation, and (c) it is stochastic with serial correlation. In case (a) the environment can be described by a (possibly) time-dependent deterministic function $W(t)$. The seasonal abundance of food is one example where such a model might be appropriate. In case (b) the environment can be modeled as a (possibly) time-dependent random variable $W(t)$ that has a distribution $\Pr\{W(t) = w\} = p(w, t)$.

In case (c), serial correlation, the probability distribution for the current state of the environment $W(t)$, depends on the past history of this variable. Weather (temperature, precipitation, and winds) is a typical example.

For cases (a) and (b), the construction of a dynamic state variable model is quite straightforward. We make the environmental parameters of the model (e.g., food supply) time-dependent, and in case (b), stochastic. Since we are not treating informational questions here, it would be assumed that the individual knows the current values of the parameters at the beginning of each time period t. Future parameter values are not known in case (b), so that the dynamic programming equation would include an expectation with respect to these values in period $t + 1$.

To model serial correlation first we consider the simplest case, in which the distribution of $W(t)$ depends only on $W(t - 1)$. Specifically, we assume that the probability distribution of $W(t)$ depends on $W(t-1)$ and on t. We denote this by

$$\Pr\{W(t) = w\} = p(w, W(t - 1), t) \tag{2.18}$$

where the function $p(w, W(t - 1), t)$ is assumed known (e.g., by estimating it from weather data). A simple example of such correlation would be

$$W(t) = \rho W(t - 1) + (1 - \rho)Q(t)$$

where $0 \leq \rho \leq 1$ and $Q(t)$ is a stationary random variable with distribution $\Pr\{Q(t) = q_j\} = p_j$. Here, today's environmental state $W(t)$ is partly determined by yesterday's, $W(t - 1)$, and partly by a random component $Q(t)$. $W(t)$ does not depend on the environmental state $W(t - 2)$ two days ago, except insofar as $W(t - 2)$ has affected $W(t - 1)$. Equation 2.18 is called a "first-order serial correlation."

To include first-order serial correlation in the model, we introduce the additional state variable $Z(t) = W(t - 1)$. We assume that the organism knows what $Z(t)$ is in period t—that is, what the environment was at time $t - 1$—but does not know what $W(t)$ will be in the current period. Its current decision can depend on $Z(t)$ but not on $W(t)$. In other words, the organism faces uncertainty about the immediate future but not about the past.

Assuming for simplicity that W (and hence Z) is a discrete random variable with possible values $w_1, w_2, \ldots, w_n$, the state dynamics of $Z(t)$ can be expressed in terms of transition probabilities $p_{ij}(t)$:

$$p_{ij}(t) = \Pr\{Z(t + 1) = w_j | Z(t) = w_i\} \tag{2.19}$$

For example, suppose Z can take three values w_1 ("good"), w_2 ("average"), and w_3 ("bad"), and suppose that p_{ij} are independent of t, with

$$
(p_{ij}) = \begin{pmatrix} .6 & .3 & .1 \\ .2 & .6 & .2 \\ .1 & .3 & .6 \end{pmatrix}
$$

(Note that the rows must sum to 1.0—why?) Here, environmental conditions tend to persist from one period to the next, and sudden transitions from good to bad, or vice versa are unlikely.

More complex models of environmental dynamics, in which the distribution of $W(t)$ depends on several values $W(t-1), W(t-2), \ldots$, would involve higher order serial correlation, but we doubt whether this level of sophistication would be worthwhile in most cases.

As an example, consider the patch-selection model of section 1.4. Now suppose that the metabolic cost $A(t)$ per period is a random variable with first-order serial correlation, as in eq. 2.19, where $Z(t) = A(t-1)$. For simplicity we assume that $A(t)$ is the same for all three patches. Now the fitness function becomes

$$
\begin{aligned}
F(x, z, t) = \ &\text{maximum expected lifetime reproduction, from} \\
&\text{period } t \text{ on, given that } X(t) = x \text{ and } Z(t) = z
\end{aligned}
$$

Now the dynamic programming equations are

$$
F(x, z, T) = \Phi(x) \tag{2.20}
$$

and

$$
F(x, z, t) = \max_i V_i(x, z, t) \tag{2.21}
$$

$$
\begin{aligned}
V_i(x, z, t) = \ &(1 - m_i)[p_i E_a\{F(x + y_i - a, a, t+1)|A(t-1) = z\} \\
&+ (1 - p_i) E_a\{F(x - a, a, t+1)|A(t-1) = z\}] \tag{2.22}
\end{aligned}
$$

(The last equation is modified for $i = 3$, the reproductive patch.) To explain this, note that the value of $A(t) = a$ is conditional on $A(t-1) = z$. Writing

$$
a_{ij}(t) = \Pr\{A(t) = a_j | A(t-1) = a_i\}
$$

we can express these expectations as

$$
E_a\{(\text{function of } a)|A(t-1) = z\} = \sum_j a_{ij}(t) \cdot (\text{function of } a_j), \text{ where } z = a_i
$$

Equations 2.21 and 2.22 assume that the patch decision is made before the individual knows the current value of a. You may wish to consider how these equations would be altered in the opposite case.

To summarize, by introducing an additional state variable $Z(t) = W(t-1)$, the case of one-period serial correlation in any environmental variable $W(t)$ can be included in a dynamic state variable model. Multiperiod correlation would require the use of additional state variables.

What difference would it make to include serial correlation, relative to a model that ignores it? In the previous patch-selection model, for example, suppose the correlation between $A(t-1)$ and $A(t)$ is positive; thus the forager anticipates higher than average costs if the previous period's costs were high. This would tend to favor building up greater energy reserves than normal in such a situation.

Which model would be superior in an empirical case? Clearly, including serial correlation may be more realistic than ignoring it—often the data will show strong serial correlation. But we advise you not to fall victim to the desire for ever-greater realism. Less realistic, simpler models often outperform more realistic, complex models. The important question is not how realistic your model assumptions are, but how well the model helps to increase your understanding of nature. Merely including environmental stochasticity may be a big advance relative to models that ignore it, or average it out; throwing in "realistic" serial correlation may not add much further insight. Such questions of model design and evaluation are discussed by Hilborn and Mangel (1997).

2.6 Errors in decisions

The fitness function $F(x, t)$ has been defined here as an individual's expected future lifetime reproduction, assuming that the individual always employs the optimal decisions. But what if the individual makes occasional (or, for that matter, regular) errors? Presumably small infrequent errors would have a minimal effect on fitness, whereas large frequent errors would have a greater impact and would be strongly selected against.

There are two ways of looking at this question. First, we can compute the modified fitness $F_{mod}(x, t)$ that corresponds to any given decision strategy $D_{mod}(x, t)$. This can be accomplished using an analogy to normal dynamic programming:

$$F_{mod}(x, t) = E\{F_{mod}(x', t + 1)\} \qquad (2.23)$$

where now $x' = X(t + 1)$ is obtained from the modified decision $D_{mod}(x, t)$. An example of this procedure was discussed in section 1.6. Then the ratio $(F(x, t) - F_{mod}(x, t))/F(x, t)$ provides a measure of the relative cost of the suboptimal strategy $D_{mod}(x, t)$ (McNamara and Houston 1986).

Second, there are many reasons for supposing that organisms are unlikely to behave exactly as predicted by a particular optimization model. Among these reasons are environmental or phenotypic variation not included in the model, lack of complete information on the part of the organism, genetic limitations, and neurological limitations. Would it be possible to include random errors in decision making directly into a dynamic state model from the outset? What would be the advantages of doing so?

Returning to the basic patch-selection model, now suppose that the organism chooses the optimal patch $i = D(x, t)$ most of the time, say with probability δ, but a wrong patch $j \neq i$ with probability $(1 - \delta)/2$. By the "optimal" patch we mean now the patch that maximizes expected future reproduction, subject to the possibility of future errors. Let $\tilde{F}(x, t)$ denote the fitness function under this assumption. Then, first

$$\tilde{F}(x, T) = \Phi(x) \tag{2.24}$$

For $t < T$, we define

$$\tilde{V}_i(x, t) = \text{fitness from choosing patch } i \text{ at time } t$$

where now fitness is calculated using $\tilde{F}(x', t + 1)$. Let $i^* = i^*(x, t)$ be the optimal patch—that is the patch that maximizes $\tilde{V}_i(x, t)$. Then, allowing for errors, we obtain

$$\tilde{F}(x, t) = \delta \tilde{V}_{i^*}(x, t) + \frac{1}{2}(1 - \delta)[\tilde{V}_{i_1}(x, t) + \tilde{V}_{i_2}(x, t)] \tag{2.25}$$

where i_1, i_2 are the other two patches.

A slightly more sophisticated, and perhaps realistic, alternative would allow the probability of choosing a wrong patch to depend on the fitness cost of doing so.

At the moment this may seem like a rather pointless exercise. However, McNamara et al. (1997) have shown that introducing random decision errors into game-theoretic dynamic models of behavior can achieve convergence in the standard iterative scheme for numerical computation of evolutionarily stable strategies (ESSs). An example is discussed in section 10.2.

2.7 Variable constraints

Although constraints on state variables have been treated as fixed constants so far in this book, there are several reasons for thinking more broadly about constraints. First, most organisms undergo growth and development. For example, the constraint $x_{\max}$ on energy reserves will typically depend on the current size of the organism. We study a model of this kind here.

Second, a constraint on a phenotypic state variable may itself be subject to selection. For example, digestive capacity may constrain the rate at which food can be assimilated. A larger digestive capacity would be beneficial in allowing the organism to increase its consumption rate but could incur increased costs of maintenance or risk of predation. If G denotes gut capacity, a dynamic state variable model could be used to calculate fitness $F_G(x, t)$ depending on G. The optimal G would maximize initial fitness $F_G(x_1, 1)$.

More broadly speaking, optimal gut capacity G could also depend on time t. For example, bar-tailed godwits (*Limosa lapponica*) apparently "dispense with parts of their 'metabolic machinery' that are not directly necessary during flight and rebuild these organs upon arrival at the migratory destination" (Piersma and Gill 1998). A model of this behavior would treat 'metabolic machinery' as a dynamic state variable, which provides a constraint on the rate of energy assimilation. Dynamic state variable models are sufficiently flexible to include such variable constraints. Next we examine one such model.

Body size as a constraint on reserves

We consider an organism that can grow and that also requires energy reserves to hedge against starvation. We introduce two state variables

$$B(t) = \text{lean body mass at the start of period } t$$
$$X(t) = \text{energy reserves at the start of period } t$$

We measure both state variables in the same units (e.g., kJ) and assume that energy reserves are constrained by body size: $X(t) \leq kB(t)$ where k is constant. For simplicity, we assume that $k = 1$, so that

$$X(t) \leq x_{\max} = B(t) \tag{2.26}$$

This may seem unnecessarily restrictive, but it merely amounts to a scaling of the variable $B(t)$ so that one unit of body size B implies one kJ of maximum reserves $x_{\max}$. For example, suppose that in reality a 100g (lean) individual has maximum energy reserves of 2000 kJ and that maximum reserves are proportional to body size for larger or smaller individuals. Then, the body-size state variable $B(t)$ is related to actual lean body mass $M(t)$ in grams by $M(t) = (100/2000)B(t) = 0.2B(t)$. Such scaling of state variables is often useful since it reduces the number of parameters needed in the model.

We also assume that the organism dies if energy reserves fall below x_{crit}, where

$$x_{\text{crit}} = k_c B(t) \tag{2.27}$$

where $k_c = \text{const} < 1$.

Reproduction occurs at time T and is given by

$$\Phi(b) = \frac{g_0 b}{1 + g_1 b} \quad \text{provided } X(T) \geq x_{\text{crit}} \tag{2.28}$$

where $B(T) = b$ and g_0, g_1 are positive constants. We also suppose that the probabilities of finding food, and of surviving depend on body size b:

$$\Pr\{\text{find food, per period}\} = \lambda(b) = \lambda_0(1 - e^{-\lambda_1 b}) \tag{2.29}$$

where λ_0, λ_1 are positive constants. If λ_1 is very large, $\lambda(b) \approx \lambda_0$—that is, the probability of finding food is independent of body size. Similarly,

$$\Pr\{\text{survive period}\} = \sigma(b) = e^{-m_0 - m_1/b} \tag{2.30}$$

where $m_0 > 0, m_1$ are constants. If $m_1 = 0$ body size has no effect on survival. Finally, we assume that

$$\text{Metabolic cost per period} = \alpha(b) = \alpha_0 b \tag{2.31}$$

where $\alpha_0 > 0$ is constant.

Food occurs in a single "bite size" Y (kJ). If successful in finding food, the individual can allocate it either to growth or to increasing its energy reserves. Therefore, the state dynamics (subject to $x_{\text{crit}} \le x \le b$) are

1. If no food is found in period t,

$$\begin{aligned} X(t+1) &= X(t) - \alpha_0 B(t) \\ B(t+1) &= B(t) \end{aligned} \tag{2.32}$$

2. If food is found and allocated to reserves,

$$\begin{aligned} X(t+1) &= X(t) - \alpha_0 B(t) + Y \\ B(t+1) &= B(t) \end{aligned} \tag{2.33}$$

3. If food is found and allocated to growth,

$$\begin{aligned} X(t+1) &= X(t) - \alpha_0 B(t) \\ B(t+1) &= B(t) + \varepsilon Y \end{aligned} \tag{2.34}$$

where ε is a measure of conversion efficiency from food to body size.

Fitness is defined as

$$\begin{aligned} F(x, b, t) = &\text{ maximum expected reproduction from } t \\ &\text{ to } T, \text{ given that } X(t) = x \text{ and } B(t) = b \end{aligned} \tag{2.35}$$

Thus,

$$F(x, b, T) = \Phi(b)$$

and for all t,

$$F(x, b, t) = 0 \quad \text{if } x < x_{\text{crit}} \tag{2.36}$$

If food is found, the fitness value of using the food for maintaining energy reserves is given by

$$V_{\text{maint}}(x, b, t) = F(x - \alpha_0 b + Y, b, t + 1) \tag{2.37}$$

and the fitness value of using it for growth is given by

$$V_{\text{grow}}(x, b, t) = F(x - \alpha_0 b, b + \varepsilon Y, t + 1) \tag{2.38}$$

We assume (fixed) constraints on body size:

$$b_{\text{min}} \leq B(t) \leq b_{\text{max}} \tag{2.39}$$

The resulting dynamic programming equation is

$$\begin{aligned}
F(x, b, t) = {} & [1 - \lambda(b)]\sigma(b)F(x - \alpha_0 b, b, t + 1) \\
& + \lambda(b)\sigma(b) \max \{V_{\text{maint}}, V_{\text{grow}}\}
\end{aligned} \tag{2.40}$$

Predictions

To illustrate the model, we used the following parameter values: $T = 10$, $k_c = 0.3$, $g_0 = 1$, $g_1 = 2$, $\lambda_1 = 10$ (so that the probability of finding food is essentially independent of body size), $m_0 = 0.04$, $m_1 = 0.4$, $\alpha_0 = 0.1$, $\varepsilon = 0.2$, $b_{\text{min}} = 3$, and $b_{\text{max}} = 20$. The parameters λ_0 and Y were varied, keeping the mean food discovered constant, $\lambda_0 Y = 0.8$.

Since $E(\text{food}^2) = \lambda_0 Y^2$,

$$\text{var (food)} = \lambda_0 Y^2 - (\lambda_0 Y)^2 = \lambda_0(1 - \lambda_0)Y^2 \tag{2.41}$$

and the coefficient of variation of food found per period is

$$CV(\text{food}) = \frac{\sqrt{\text{var(food)}}}{\lambda_0 Y} = \sqrt{\frac{1 - \lambda_0}{\lambda_0}} \tag{2.42}$$

Thus food discoveries are highly variable if $\lambda_0 \ll 1$. A variable food supply means that starvation risk is high unless reserves are maintained at a high level. Table 2.1 shows the number of (discrete) states and times for which the optimal behavior is to use food for growth. When food is found in nearly every period ($\lambda_0 = 0.92$), about 40% of the combinations of states and times lead to a prediction that the optimal behavior is to use the food for growth.

Table 2.1 Effects of variability in finding food on the use of that food

| λ_0 | CV(food) | Number of states and times for which the optimal behavior is: | |
		Eat	Grow
0.2	2.0	1075	302
0.4	1.22	1038	339
0.6	0.82	998	379
0.8	0.50	885	492
0.92	0.30	800	577

On the other hand, when food is rarely found ($\lambda_0 = 0.2$), it is optimal in only about 22% of the combinations to use the food for growth. This result accords with our intuition: environments that are more variable are inherently more risky because of the chance of starving to death, and the result is that there are fewer states for which it is optimal to use the food for growth.

Now, however, imagine the following experiment. Individuals from environments characterized by different levels of variability are taken to the laboratory and grown under conditions in which they receive food every period. To begin, we assume that regardless of source environment, all individuals receive the same amount of food Y_f per period. To predict the trajectory of body size, we used Monte Carlo forward iteration; see section 2.3.

Figure 2.3 shows simulated body size $B(t)$ for various values of CV(food). In agreement with intuition, the model predicts that individuals that use behavior appropriate to highly variable food supplies grow less than those from a less variable environment. The laboratory experiment could be used to test this qualitative prediction.

Now envision a variant of this experiment in which organisms are given the same food Y_f as in their source environment but are given it in every period. Thus, for example, an organism from the least stable environment (where $Y = 4$ and $\lambda_0 = 0.2$) receives $Y_f = 4$ in each period, and an organism from the most stable environment (where $Y = 0.87$ and $\lambda_0 = 0.92$) receives $Y_f = 0.87$ in every period. In this case, we obtain a very different and apparently contradictory result (fig. 2.4). In this case, organisms from the most variable environment grow to the largest sizes, in apparent contradiction to the previous results. But the contradiction is only apparent. In the laboratory, individuals from the highly variable environment receive food $Y = 4$ in every period. Even though they usually favor reserve maintenance over growth, these individuals quickly reach reserve saturation $x = x_{\max} = B(t)$ and then use food largely for growth. Individuals from more stable environments receive less food and grow less.

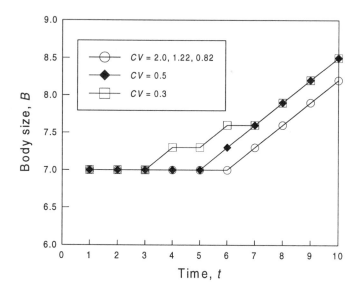

Figure 2.3 The trajectories of body size for individuals in which the coefficient of variation of food availability is 0.3, 0.5, 0.82, 1.22, or 2.0; here $Y_f = 1.5$. Note that although the allocation behaviors are different, according to the solution of the dynamic programming equation (table 2.1), many of the subtle differences do not appear in the forward iteration.

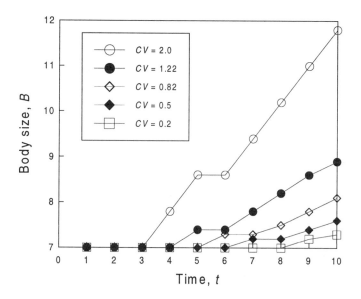

Figure 2.4 Predictions of body size when organisms receive, in each period, the amount of food present in their source environment.

The simple model shows how what we've previously treated as a constraint can be viewed as a state variable. It also has implications for the way we think about testing optimality models and shows the care that must be taken when we try to mimic nature in the laboratory if we are to obtain sensible results. These questions are discussed further in chapter 3.

2.8 Functional forms

When developing a dynamic state variable model, you often need to assume a functional form $y = f(x)$ with certain features. The function $f(x)$ will involve a number of parameters $a, b, \ldots$ (usually two or three are sufficient); we write $y = f(x; a, b, \ldots)$. If data (x_k, y_k) are available, the parameter values can be estimated, for example by minimum least squares or maximum likelihood (Hilborn and Mangel 1997).

But how does one choose the functional form $f(x; a, b, \ldots)$ to begin with? There are at least three approaches: (1) use the data as a guide; (2) use common sense ideas about the underlying biology; or (3) construct a separate model of the biological processes involved. By itself, method 1 is not very satisfying from a scientific point of view—it says nothing about why the data might look the way they do. (The term *curve fitting* is often used pejoratively to describe this mindless approach.) Method 3 is perhaps the ideal and should probably be considered at some stage. Constructing a detailed separate model for $f(x; a, b, \ldots)$, however, may involve a lot of work, which may not be worth the trouble initially. On the other hand, we have often had the experience that constructing a relatively simple model greatly aided our understanding of the situation. For example, in constructing a model of the vertical migration of planktivores, Clark and Levy (1988) needed a function relating the foraging rate of a marine predator to light intensity and prey density. A simple search model was developed, which by itself helped to explain some perplexing observations.

To illustrate this procedure, we consider the relationship between the consumption rate by an individual predator and the density of food, which is called the **functional response** (Solomon 1949; Holling 1959; Begon et al. 1990); see fig. 2.5. The simplest assumption is that the consumption rate R is proportional to the food density D, at least until some saturation level $R_{\max}$ is reached at density $D_{\max}$. Then we have (fig. 2.5)

$$R(D) = \begin{cases} cD & \text{if } D < D_{\max} \\ R_{\max} & \text{if } D \geq D_{\max} \end{cases} \tag{2.43}$$

How might we determine the constant c? Since $R(D_{\max}) = R_{\max} = cD_{\max}$, we find that $c = R_{\max}/D_{\max}$. This is called a Type 1 functional response; according to it, feeding rate increases linearly with prey density until a plateau is reached. (Begon et al. 1990, fig. 9.8, give an example.) However, we

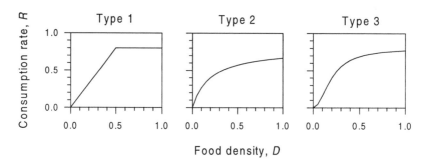

Figure 2.5 Functional response curves of Types 1, 2, and 3.

expect that as food density increases, the feeding rate may not increase proportionately—that is, feeding rate may depend nonlinearly on food density but still saturate. To create a model for this by the method of thinking about it, we begin with the dimensionless variable $x = D/D_{max}$ and recall that a simple function which saturates as x increases is $f(x) = x/(x + x_0)$, where x_0 is a parameter. Since $f(x_0) = 1/2$, we interpret x_0 as the "50% level" of food. Note too that if $x \ll x_0$, then $f(x) \sim x$ (i.e., proportional to x) and if $x \gg x_0$, then $f(x) \sim 1$ (i.e., constant). We encourage you to sketch or plot $f(x)$, once you have picked x_0. Now we return to unscaled variables, multiply $f(x)$ by R_{max}, and set

$$R(D) = \frac{D/D_{max}}{D/D_{max} + D_0/D_{max}} = \frac{R_{max}D}{D + D_0} \tag{2.44}$$

This is called a Type 2 functional response. It is often used to model obligate predators or specialists because when the density of prey is low, the consumption rate is proportional to the prey density, as if the predators had no choice but to consume this particular prey type. Begon et al. (1990, fig. 9.7) provide examples of the Type 2 functional response.

On the other hand, a generalist predator would switch to a different prey when the one that we are considering becomes too low in density. Hence, we would expect that the consumption rate declines rapidly when the density becomes small but still saturates at high densities. To capture this effect, recall that

$$f(x) = \frac{x^2}{x^2 + x_0^2}$$

has a sigmoidal shape and captures the effect that we are interested in because when $x < 1$, $x^2 < x$. Note that we kept the same interpretation for x_0 as before by squaring it.

In this case, the resulting dimensional functional response is given by

$$R(D) = \frac{R_{max}D^2}{D^2 + D_0^2} \tag{2.45}$$

This is called a Type 3 functional response. The appendix to chapter 1 and section 9.1 contain process models for functional responses.

Method 2 is often used in situations where Method 3 would be too time-consuming. We briefly discuss one example.

Parental care

Suppose that one is developing a dynamic state variable model, one component of which is parental care of offspring. Thus one needs a functional form relating parental effort e to offspring fitness $f(e)$. Presumably $f(e)$ is an increasing function of effort e, which is concave downward, and has an asymptotic limit.

In a simple graphical analysis, a general curve may be sufficient. But a dynamic state variable model usually incorporates actual numerical values. This requires that units of effort e be specified and that fitness f be given operational meaning. For example, e could be defined as the number of days during which the parent continues to feed its offspring, and fitness f might be specified as the probability that the offspring survives to reach sexual maturity. Then, in principle, a detailed process model relating f to e for a given species could be developed. Experimental manipulation could also be used to estimate this relationship. An alternative approach (Method 2) would be to use a particular functional form and to specify the parameters e_{max}, f_{max}, plus a shape parameter b.

To obtain such a functional form, we work with a dimensionless variable $x = e/e_{max}$ and look for a function $f_1(x)$ that the properties that

$$f_1(0) = 0, \quad f_1(1) = 1, \quad f_1 \text{is increasing}, \quad f_1'(0) = b \ (b > 1) \tag{2.46}$$

After some trial and error, we came up with

$$f_1(x) = \frac{bx}{1 + (b-1)x} \tag{2.47}$$

You should graph this function and check that it meets the required conditions.

Finally, we scale e and f to get the desired function

$$f(e) = \frac{be/e_{max}}{1 + (b-1)e/e_{max}} f_{max} \tag{2.48}$$

This function involves three parameters, e_{max}, f_{max}, and b. Unless actual data are available, values for these parameters can be determined on the basis of the biology of the animal being modeled.

If you expect to model as a regular activity, it will be worth your while to develop the skill to generate a variety of functional forms. Computer software such as MathCad or Mathematica can be helpful in rapidly depicting different graphs.

3

Using the Model

All models are wrong, but some models are less wrong than others.
— with apologies to George Orwell

3.1 Levels of detail in model predictions

The main purpose of most of the models discussed in this book is the understanding of adaptation, which G.C. Williams has called "a phenomenon of pervasive importance in biology" (Williams 1966, p. 5). In this quest, the modeler will often be studying some behavioral observation that may not previously have been considered at all as an adaptation. Alternatively, previous explanations may have been based on now discredited principles such as group selection. The modeler has formulated a novel, potential, adaptive explanation. At a purely verbal level, the new hypothesis may seem credible but leaving the subject at this level invites the criticism that the explanation is no more than a "just-so story" (Gould and Lewontin 1979).

Developing a rigorous optimization model puts the verbal hypothesis to the test: can the observed behavior be predicted from the adaptationist hypothesis, in quantitative terms, based on known, measured parameter values, and including important trade-offs? Answering this question may be far from straightforward. First, the necessary data may not be available. Second, the appropriate type of model may not be evident. Third, how much detail to include in the model may not be clear. Fourth, different hypotheses may lead to different models, which need to be evaluated.

Evaluating a model proceeds through several stages, each of which may lead to modification or outright rejection of the model. We identify four such phases: general workability, qualitative testing, quantitative testing, and generating and testing novel predictions. In the first stage, the question is whether the proposed hypothesis is actually capable of explaining the observations. Although accurate estimates of model parameters may not be available, the model should indicate whether the observations are consistent with the hypothesis under reasonable assumptions about the parameters. If not, the hypothesis may be rejected, modified, or represented in a different model.

71

On the other hand, if the model shows promise, work proceeds to more detailed parameter estimation, model development, and testing. In this chapter we discuss the questions of (a) confronting models with data and (b) generating testable predictions from dynamic state variable models. We discuss these topics in this order because we consider that the interplay of models and data is of primary importance.

3.2 Comparing models and data

After reading the previous two chapters, you should recognize the fundamental and key notion about dynamic state variable modeling: the dynamic programming equation is the model. Until you have written it down, thought about parameters, and programmed it, you have not constructed a model. Since we are using mathematics and computer programs to understand biology (in contrast to those who use biology to motivate mathematical studies or computer algorithms), comparison of models and experimental or observational data is an essential component of the research program. In this section, we discuss some of the principles and methodological considerations associated with such comparisons. We delay the discussion of particular computational methods (see, e.g., Hilborn and Mangel 1997) until they are used in the study of particular systems.

At the outset, we separate **statistical** and **mechanistic** models (Mangel et al. 1998). Statistical models are those that arise in the analysis of data (e.g., regression, ANOVA, etc). They are used to make inferences about properties of the data. On the other hand, mechanistic models, which are what we consider in this book, may lead to predictions that disagree with the data. In such a case, one must rethink the logic of the situation.

When a model can fail

Models are tools that allow us to go beyond the data (i.e., create theory) for making predictions. As such, to use models, we make assumptions that are uncertain or difficult to test. This will always be true, and one must sort out the useful predictions (e.g., those that are relatively robust to uncertain assumptions) for comparison with the natural world.

A model can fail in the obvious way, if it makes predictions that are completely discordant with the data. However, a model can fail in a more insidious way: if a model leads to "exactly what we expected" without further understanding or prediction, then it has also failed. The reason is that if the results were expected and there are no new, model-specific predictions, then you understood the scientific question before building the model. It was superfluous.

A good model should (1) explain phenomena at a level not previously understood and (2) suggest experiments or observations that have not yet been conducted. If one is particularly lucky, then, after the model has been developed and analyzed, it may be possible to understand the data without recourse to the model. This happened to us, for example, in the study of insect oviposition behavior (chapter 4) and vertical migration of fish (Clark and Levy 1988); also see Mangel and Clark (1988).

For individuals new to modeling, we suggest that you begin with either an empirical observation that is problematic or a hypothesis about empirical observations. Use these as motivations for the development of the model and challenges to the model. In addition, as we discuss here in more detail, there is a great advantage in having more than one model as you think about the data.

Variation between model predictions and observations

The models that we describe in this book are almost all optimality models, and it is important to recognize that variation is consistent with optimality models.

Indeed, optimality models cannot predict how much variation is consistent with them because variation is genetically determined. In addition, selection for optimal behavior is weakest near the optimum, so that even the most rabid adaptationist should not expect that individuals or populations will achieve the optimum exactly as predicted by a given model. Variation from the predictions of optimality models has often been used by opponents as demonstration that the models have failed. What is more important than the variation in behavior, however, is the fitness of the associated variation. But it is important to remember that optimality models with fixed parameters cannot suggest how much variability is consistent with local optimality.

Quantitative comparison of models and data

There are quantitative methods for determining the best parameter estimates and comparing models with data (see Hilborn and Mangel 1997 for an introduction to this literature), and we discuss some of these in subsequent chapters. In general, different kinds of statistical methods are used when the models are nested (i.e., a more complicated model becomes the simpler model when various parameters are set equal to zero) than if they are not nested.

However, when used with a single model and data set, the implicit assumption of this approach is that (1) statistical quantitative fit indicates that the model is "correct" and (2) lack-of-fit indicates that the model is "wrong." Let us deal with the second point first. Statistical inaccuracy can be caused by variability around the average behavior that is greater than the acceptable level of the statistical test, or by systematic deviations between the predicted values and average observed behavior.

When models involve a number of parameters, as is usually the case with dynamic state variable models, it is likely that individuals will differ with respect to the parameter values. In itself, this can lead to scatter around the predicted relationship and to statistical lack of fit if the variation in parameter values is great.

Without information on variation in parameter values among individuals or observations, it is impossible to conclude a priori how much variation in behavior is consistent with the optimality model. Consequently, statistical tests of the quantitative fit of deterministic predictions to observed data are not fair tests of the accuracy of the model, regardless of the outcome.

Systematic differences between model predictions and average behavior also make quantitative tests difficult to interpret. This type of lack-of-fit can reflect the fact that the parameter estimates are incorrect, that the model is missing an important qualitative feature, that individuals are not behaving optimally, or a combination of these factors.

The critical point is that in any case of lack-of-fit, it will be uncertain what actually causes the lack-of-fit. Errors in parameter estimates are more likely to cause deviations between observed and predicted behaviors than a fortuitous, quantitative match. Thus it will always be possible that better parameter estimates or a qualitatively different model would lead to quantitative accuracy. This is a poorly appreciated point in the generally vituperative and wasteful debate on the usefulness of optimality models in biology.

Indeed, a common reaction to a model (optimality or otherwise) is that it is "not realistic"—that one cannot model a system until everything is known about it. But then, why bother? There may be masses of data and measurements, but often nobody has asked, "what are the important parameters as far as the viability of the population is concerned" or "what drives the system?" Models are necessary to begin addressing such questions.

Thus, statistical tests of quantitative fit are often bound to fail because model predictions are deterministic, but behavior is genotypically and phenotypically variable, and because errors in parameter estimation are likely to lead to deviations between predicted and observed behavior. Moreover, lack of statistical fit gives little clue to the reason that a model does not fit. Finally, a quantitatively accurate statistical fit may be good luck rather than good science.

In summary, the lack of a "good" (i.e., statistically accurate) fit between the numerical predictions of an optimality model and the observations need not be grounds for rejecting the model. However, if the model predicts values that are consistently at odds with the data, then clearly the model is suspect. Then further consideration may suggest that in fact the model fails to represent reality in some way. Indeed, that is the way it should be: in almost every situation, our own efforts early in a project often founder in this way, forcing us to abandon one model, or one component of a model, and develop a more convincing alternative. Although such failed models seldom see publication, they are extremely valuable in advancing one's understanding of nature, which is the point of the effort. Indeed, it is not uncommon to do ten times as much modeling and computation as ever gets published—at least, this is our experience!

If a given model seems reasonable, it may be possible to tune certain of its parameters whose values are not accurately known, so as to yield better quantitative predictions. It is essential to candidly admit to this tuning to avoid the charge of falsifying the input data. There are two advantages to such honest tuning. First, if successful, it indicates that at least the model is capable of making good quantitative predictions. Second, the tuned parameter value that results in the best quantitative predictions is itself a prediction—an inverse prediction of the parameter value (or range) based on the requirement that the predictions of the model match the data. Assuming that experiments can be devised to measure the unknown parameter, the inverse prediction can be tested. If the tuned value is outside the range of measured (or realistic) values, this constitutes evidence of model failure.

For example, many dynamic state variable models require an assumption of fairly substantial predation risk (examples occur later). The modeler may be criticized on the grounds that such levels of predation are unrealistic. How often does one observe a predator killing a small bird, for example? As a first check, one can estimate the abundance of predators and their total food requirements and figure out the corresponding rate of predation for the prey species in question. This exercise often results in a much larger estimate of predation risk than casual observation might suggest. It is also worth noting that individuals with suboptimal states or behavior (e.g., too fat to maneuver quickly or too nonvigilant) may be highly at risk relative to the average member of the population. Thus potential predation risk, which is what is used in the model, may be much greater than the measured predation rate (Ydenberg 1998).

Qualitative tests of models and data

In many situations, qualitative tests are comparatively more useful, more appropriate, and relatively immune to the problems associated with quantitative tests (Rosenzweig and Abramsky 1997, p. 752). First, qualitative tests are much more likely to be robust to the assumptions about optimality and terminal fitness. In particular, qualitative understanding requires only that the model correctly predicts the direction of the behavioral response; that is, approximates the general shape of the fitness surface.

Second, if one views qualitative predictions as hypotheses, then the tests are of those hypotheses, rather than of the models (or the modeling approach), and an empirical test becomes meaningful whether the predictions are met or fail. If genetic or other constraints prevent organisms from behaving in the direction predicted by the model, then the hypothesis would be rejected and the unsuccessful test would provide both valuable information and impetus for developing alternative hypotheses. Successful tests provide support for specific hypotheses generated by dynamic state variable models.

Third, qualitative tests are generally less sensitive to unexplained variation in behavior and to uncertainty in parameter estimation than tests of quantitative predictions.

The main challenge is to use dynamic state variable models to identify the key factors that affect the trade-offs that determine behavior and to derive testable predictions (qualitative or quantitative) from these models.

A procedure for comparing models and data

Now we outline a guide for comparing your model and data. The particulars in any specific case will require modification of these general guidelines.

1. Identify robust predictions about the way one or more factors will influence behavior. In some cases, this will be easy; the predicted relationship will be a monotonic function, independent of other parameter values. In other cases, the relationship will not be monotonic, or its direction will depend on the values of other parameters in the model. Thus it may be helpful to obtain better quantitative estimates of parameter values. In either case, the key is to identify and focus on testing qualitative predictions that are appropriate for the range of conditions observed in nature and that are robust to variation in parameters that are variable and difficult to estimate. This is very much a kind of sensitivity analysis (see section 3.4).

2. Robust qualitative predictions are best and most rigorously tested by experimentally manipulating the factor(s) predicted to affect behavior (with appropriate controls). If an experimental manipulation is impossible, then statistically relating variation in behavior to naturally occurring variation in the factor can provide a corroborative test of the prediction. However, observational tests of predictions are prone to alternative explanations. Only experiments can demonstrate a causal link between behavior and the factor as predicted by the model (e.g., Hairston 1989; Krebs and Davies 1993)—and indeed one definition of science is that experimentation is the sole arbiter of truth (Feynman 1965). Other observational comparative approaches (e.g., Harvey and Pagel 1991) can also provide corroborative evidence, often at a scale impossible to test experimentally, and should be viewed as complementary to more rigorous experimental approaches (e.g., Krebs and Davies 1993).

3. Experimental and observational tests of qualitative predictions should be regarded as tests of hypotheses rather than tests of optimality theory or of dynamic state variable models. A statistically significant trend in the predicted direction provides support for the hypothesis. A negative result yields insight and should stimulate the development of alternative hypotheses.

4. This approach can be further improved by focusing on testing the qualitative predictions of alternative models (see Hilborn and Mangel 1997). The idea is to identify conditions in which models based on different biological assumptions make qualitatively different predictions. Arguably, this approach has the potential to yield the greatest insight into the processes and factors that affect adaptive behavior. The challenge for both theoreticians and empiricists is to find and exploit conditions in which mechanistically different models make qualitatively different predictions.

5. Finally, we note that testing the assumptions of optimality models, investigating nonadaptive explanations of the qualitative aspects of behavior,

and studying the underlying genetics and physiology of behavior are equally valid and useful approaches. Overall, a research program that integrates several or all of these to study a single behavior or suite of behaviors may be the most productive approach.

Many behaviors that were considered incomprehensible in adaptive terms (and thus attributed to other causes, including genetic or physiological constraints or selection at the group level) have recently been understood as bona fide adaptations. Examples include superparasitism in ovipositing insects (Mangel and Roitberg 1988; see chapter 4), prefledging mass decline in birds (Clark and Ydenberg 1990a,b) and diadromy in fish (Gross 1987). The increasing use and sophistication of optimality models, including game-theoretic models, has contributed strongly to this deeper understanding of the nature of ecological adaptation.

A methodology for quantitative testing

There will be cases in which one wants to conduct a quantitative test. For example, if two or three substantially different models all give the same qualitative prediction over the reasonable range of parameter values, then it is not possible to differentiate among them by a qualitative test. In that case, we view the situation as a competition between the different models, and the data arbitrate the results of the competition (Hilborn and Mangel 1997).

One of the difficulties with quantitative tests is that we almost never know the sources of uncertainty or statistical error in behavioral models. Making ad hoc assumptions is relatively easy, but is also quite dangerous. For that reason, computer-intensive methods (Efron and Tibshirani 1993) are particularly useful. We suggest the following procedure:

1. Generate many replicates of your data by resampling the data with replacement. These are usually called "bootstrap" datasets and the idea is that, based on the assumption that the original data are representative of the system, you have in some sense replicated the experiment many times.

2. At first, confront each model with the data. Suppose that there are N data points and that model i has p_i parameters that are estimated by comparison with the data. Given the parameters, one can compare predictions of the model with the data, for example by using forward iteration. These predictions lead to a **sum-of-squares deviation** between the model and data. For example, if $X_{obs}(t)$ were the observed value of the state variable (never mind that this might be hard to do) at time t and $X_{ik}(t)$ were the simulated values assuming model i and forward iteration replicate k ($k = 1, \ldots, K$), the sum-of-squares deviation is

$$SSQ_i = \frac{1}{K} \sum_{k=0}^{K} \sum_{t=0}^{T} [X_{obs}(t) - X_{ik}(t)]^2 \qquad (3.1)$$

The same kind of measure can be constructed if the observed and pre-dicted quantities are behaviors—for example, clutches laid, food delivered to offspring, or foraging effort during the day.

3. Compare the models by choosing the one that has the smallest value of $SSQ_i/(N - 2p_i)$. This quantity (Mallows 1973; Efron and Tibshirani 1993; Hilborn and Mangel 1997), penalizes models with many parameters in the same way that the denominator $N - 1$ is used to estimate variance.

In the following chapters, we shall apply many of these ideas concerning qualitative and quantitative tests of models.

3.3 Obtaining and analyzing predictions

Numerical solution of the dynamic programming equation yields two state- and time-dependent functions, the fitness function $F(x, t)$ and the decision function $D(x, t)$. The fitness function represents expected future reproduction, or "reproductive value" in the terminology of life-history theory. Usually fitness is not an observation that one is attempting to explain, although measurement of fitness and its dependence on an organism's state is possible in principle and sometimes in practice. But even if you don't plan to measure fitness, we recommend that you examine the fitness function as an aid to checking the correctness of your calculations. Occasionally, you may find that the computer is producing obviously outrageous values for $F(x, t)$—a clear sign that your code has an error (which would be embarassing to learn about later). You should also check that $F(x, t)$ depends in the expected way on the state x and time t. If not, either your code is in error, or there are logical implications of your model that you do not understand.

The decision function $D(x, t)$ provides immediately testable predictions, namely, the dependence of behavior on time and on the individual's current state. Indeed, the general statement that behavior is often state-dependent is itself testable, and this assertion has recently been established empirically in many organisms, for example Minkenberg et al. (1992); Bull et al. (1996). Qualitative and quantitative predictions about behavioral dependence on time or state in specific organisms under specific circumstances are readily testable in the field and in the laboratory.

Graphical representation

Well designed graphics are usually easier to understand than tables of values. Graphs of the decision function $D(x, t)$ can range from simple to gothically complex. In the basic patch-selection model, for example, x is a single state variable, and only three decisions (patch numbers) are considered. The graph of $D(x, t)$ is correspondingly simple (fig. 1.2). In other models in which the decision variable $D = D(x, t)$ is continuous, the graph becomes a surface in three dimensions. Computer graphics makes the rendering of 3-D graphics easy, although care is needed to get the most easily understood picture. The qualitative features of $D(x,t)$ are apparent from a well-designed graph. What if the state variable is two (or more) dimensional? For example, the graph of $D = D(x,y,t)$ would be a hypersurface in four dimensions! Don't despair—this can also be depicted in an understandable way. To do this, choose a representative set of t-values, $t_1 < t_2 < \ldots < t_n$, and draw n 3-dimensional graphs $D = D(x,y,t_i)$ arranged on the page. Clark and Harvell (1992) used this method to show five-dimensional decision functions. Beyond that it gets a little complicated!

Similar methods can be used to display other predictions such as parameter sensitivity (see section 3.4). Complicated graphics are easily overdone, however. Most people readily comprehend two-dimensional, plane graphs but are less familiar with 3-D (not to mention 5-D) graphics. Often even a three-dimensional graph $D = D(x,t)$ can be depicted as a set of two-dimensional graphs $D = D(x,t_i)$ for $t_1 < t_2 < \ldots$. Alternatively, contour plots (handled routinely by most graphics packages) may be easier to understand. It depends on circumstances. Be thoughtful and creative in choosing graphical representation for your model's predictions.

Uses of forward iteration or simulation

An example of a prediction obtained from simulation of state trajectories (section 2.3) is the timing of migration. In a model of migration, the fitness function $F(x,z,t)$ depends on the organism's internal state $X(t) = x$—for example, energy reserves—and on its current geographical location $Z(t) = z$. The simulated state trajectory $(X(t), Z(t))$, in particular, yields a prediction of the migration schedule $Z(t)$. Stochastic environmental conditions such as weather and winds imply that $Z(t)$ is also stochastic and year-to-year variation is caused by the environment. Then the predicted mean and variation in the migration schedule $Z(t)$ can be compared with field observations. In chapter 6 we discuss one such example in detail, in which interannual variation is mainly weather-induced. Predictions are of the mean dates of peak population numbers at each location z and the variance about these mean dates in different years.

In other models, such as the oviposition models discussed in chapter 4, variation in individuals' states $X(t)$ at any given time and over time results from random encounters with food items or patches, hosts for egg-laying, mates, or predators. In other words, these models predict event-triggered phenotypic variation within a population. Other sources of variation such as individual genetic differences or spatial environmental variation can also be analyzed using state variable models. This is best treated by sensitivity analysis, which we discuss next.

3.4 Sensitivity analysis

The term **sensitivity analysis** refers to the effect that changes in a given parameter value have on selected predictions of the model. This has two important applications. The first is to reveal which parameters are the most important, and in particular, whether model predictions depend strongly on poorly known parameters (or other aspects of the model). It is typical that the process of developing and testing a model leads to changes in your understanding of the relative importance of different aspects of the biology. Indeed, this is one of the main purposes of modeling.

The results of a sensitivity analysis are often presented in terms of elasticities. For example, if a 5% increase in the value of a certain parameter (x) results in a 2% decrease in a certain output value (y), we would say that the elasticity of y relative to x is $(-0.02)/0.05 = -0.4$. Elasticities have the advantage of being dimensionless; however, sensitivity can also be expressed dimensionally as $\Delta y/\Delta x$, with dimensions (units of y) $\times$ (units of $x)^{-1}$. In both cases, Δx is assumed small relative to x, and Δy is small relative to y.

A second use of sensitivity analysis is related to model predictions. Thus, a certain model parameter may differ from one location to another, and the model may predict qualitatively or quantitatively different results for individuals at different locations. If these predictions are supported by the data, confidence in the generality of the model is increased, and vice versa. Similarly, experimental manipulation may be used to deliberately alter the value of some parameter. The model's predictions of the direction and magnitude of one or more output values is directly testable. As we noted before, such controlled experimental testing of models is the essence of good science.

Sensitivity testing—that is, evaluating the implications of changes in the model—can help in understanding biological diversity at various scales. Different individuals in a given population may encounter different microhabitats, resulting in different behavior. This in turn can affect the individual's growth, development, and reproductive success. On a broader scale, different populations of a certain species may inhabit quite different environments; sensitivity analysis of a dynamic state variable model can help in disentangling the multiple effects that environment has on life history, behavior, and reproduction.

Environmental impact assessment

A particular application of sensitivity analysis pertains to assessing the impact of environmental changes such as loss or degradation of habitat or changes in climatic conditions. A dynamic state variable model can be used to estimate the impact of such changes on the average lifetime reproductive success of individuals in a particular population. This result does not provide a direct estimate of the expected change in population size, other than to predict, for example, that a given environmental change could lead to a rapid initial decrease in the population. In any event, predicting the long-term population consequences of various environmental changes is notoriously difficult because it requires a detailed understanding of density-dependent effects at all stages of the species' life history and annual cycle (Sutherland 1996).

A dynamic state variable model can be used to assess both short-term and long-term effects of environmental change. In the short term, it is assumed that a new behavioral strategy appropriate for the altered environment does not evolve, but rather that the original strategy continues to be employed ("hardwired" behavior). If $D(x, t)$ denotes the originally optimal strategy (i.e., decision function), let $F_{new}(x, t)$ denote the (nonoptimal) fitness function of an individual that uses strategy $D(x, t)$ in the altered environment. The following backward iteration equation can be used to compute F_{new}:

$$F_{new}(x, t) = E\{F_{new}(x'_D, t + 1)\} \qquad (3.2)$$

where E denotes expectation and $x'_D = X(t + 1)$ is the individual's state in period $t+1$, using strategy $D = D(x, t)$. Note the absence of any maximization expression on the right side of this equation, reflecting the assumption that the strategy D is hardwired in the short term. The computer code used for the original model needs only minor modification to calculate the new fitness function $F_{new}(x, t)$, using an equation like eq. 3.2.

The ratio $F_{new}(x_1, 1)/F(x_1, 1)$ gives an estimate of the relative change in fitness that would result from the environmental change under consideration. An example of this method appears in chapter 8. (There is no a priori reason to expect that this ratio will be less than 1, by the way—an environmental change could possibly enhance fitness.)

The long-term implications of environmental change can be calculated from the original dynamic programming code by altering the parameters and calculating the new fitness function $F'(x, t)$. As before, the ratio $F'(x_1, 1)/F(x_1, 1)$ represents the relative long-term change in fitness.

3.5 Suboptimality analysis

A criticism of dynamic state variable models may be that the predicted behavioral strategies are too complex to ever evolve by natural selection (Mangel 1990b, 1991). **Rules of thumb** are simple behavioral strategies that seem biologically realistic; the relative degree of suboptimality of any rule-of-thumb strategy can be computed by using a dynamic state variable model. The procedure, which is the same as that discussed in the previous subsection, can also be used to compute the loss of fitness that results from any variation from optimal behavior (McNamara and Houston 1986).

Specifically, if $D(x, t)$ denotes a given rule of thumb or other suboptimal strategy, then the fitness function $F_D(x, t)$ can be computed from an equation analogous to eq. 3.2. Then the ratio $[F(x_1, 1) - F_D(x_1, 1)]/F(x_1, 1)$ is a measure of the cost of the suboptimal strategy D.

4

Oviposition Behavior of Insect Parasitoids

Techniques used in chapter 4

Fitness measures (4.1)
Static optimality models (4.2)
Rate-maximizing models (4.2)
State-independent dynamic models (4.3)
State-dependent dynamic models (4.4)
Multiple state variables (4.5)
Qualitative testing of models (4.2–4.5)

In this chapter we describe the application of dynamic state variable models to insect parasitoids (which we describe in more detail below). This has been an especially fruitful area of application, in which dynamic state variable models have yielded new insights and suggested a number of new experiments.

4.1 Parasitoid life histories

Parasitoid life styles are manifold (Godfray 1994), but for simplicity, one can envision them as follows (Price 1980). Adults (who may live just a few hours up to many weeks) are free ranging. Eggs are laid on or inside various life stages of other insects, ranging from eggs to adults. Upon hatching, the offspring use the body fluids and tissues of the host as resources for growth. Offspring may complete development in the host or may exit and pupate elsewhere (for example, in the ground); offspring development generally results in the death of the host. Thus the key notion is that offspring are "trapped" in the host, even though the adults are mobile. This makes parasitoids easier to model than moths or butterflies, in which offspring can move (e.g., Mangel and Roitberg 1993; Roitberg and Mangel 1993). Parasitoids are often used for biological control of pest insects. However, insects with parasitoidlike life styles such as tephritid fruit flies (e.g., Mediterranean fruit fly, apple maggot) can often be pests themselves.

David Lack's ideas on clutch size and their generalization

The great evolutionary biologist David Lack introduced notions about clutch size in birds (Lack 1946, 1947, 1948a,b; Monaghan and Nager 1997) that remain influential in evolutionary biology 50 years later (e.g., Godfray et al. 1991; Risch et al. 1995). Lack's approach is based on two fundamental observations. First, individuals can generally lay more eggs than they do. Second, in general there is density-dependent competition for resources among the offspring—laying more eggs means that the share of resources per offspring is smaller because competition will be more intense. This implies that there is a trade-off between eggs laid and some measure of reproductive success (we discuss these measures in more detail later) and that there will be an "optimal clutch size."

We illustrate these ideas for insects, using data collected by Rosenheim and Rosen (1991), who studied the parasitoid *Aphytis lingnanensis*, which attacks the armored scale insect, a worldwide pest of citrus. They found that the size of an emerging daughter, measured by hind tibia length, depends upon the clutch laid by the mother. In particular, if $S(c)$ is the average size of a daughter from a clutch of size c

$$S(c) = 0.245 - 0.0223(c - 1) \tag{4.1}$$

Furthermore, the number of eggs that a female parasitoid can lay depends upon her size and thus upon the clutch from which she emerged. If $E(S(c))$ denotes the number of eggs, Rosenheim and Rosen found that

$$E(S(c)) = \max\{181.8S(c) - 26.7, 0\} \tag{4.2}$$

where $\max\{A, 0\} = A$ if $A > 0$ and 0 otherwise. We combine these equations to create a measure of fitness, potential granddaughters, for the ovipositing mother as a function of the clutch that she lays, that is, each daughter from a clutch of size c has size $S(c)$ and the potential to lay $E(S(c))$ daughters herself, so that the potential number of grandchildren is $cE(S(c))$; fig. 4.1. In subsequent analyses, we shall use $f(c) = cE(S(c))$ as the increment in fitness obtained from oviposition of a clutch of size c in a single host. It is possible, of course, that the number of eggs also affects the survival of offspring in a host (in addition to their size at emergence). We defer discussion of that case (see Rosenheim and Rosen 1991).

Examining fig. 4.1 is instructive. Although, based on eq. 4.2, the mother can lay as many as 11 eggs and still have offspring emerge, we see that hosts that receive more than 6 eggs give rise to daughters who are so small that they are functionally infertile. There is an optimum clutch size for a single host,

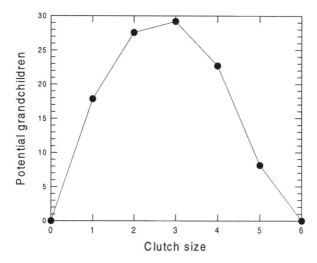

Figure 4.1 The potential number of grandchildren emerging from a host as a function of the clutch laid in the host for the data on *Aphytis* given by Rosenheim and Rosen (1991).

which we call the **single host maximum (SHM)**. Thus a clutch of 3 eggs gives fitness of 29.2 potential grandchildren, but clutches of 2 eggs give fitness 27.6 potential grandchildren, nearly as much. This simple observation leads us to recognize the need for dynamic state variable models. Imagine a female who has exactly 6 eggs left. She could put three eggs into each of two hosts and obtain fitness of 58.4 potential grandchildren. Alternatively, she could put two eggs into each of three hosts and obtain fitness of 82.8 potential grandchildren. On the other hand, if she were guaranteed to encounter six hosts, then she could put one egg into each of them and thus obtain fitness of 17.1 potential grandchildren per host, or total fitness of 102.6 potential grandchildren. Thus our predictions of what she does can depend upon the number of eggs that she currently holds (**egg complement**), the chance of encountering hosts, and the chance of mortality. It is the role of these factors that we want to sort out.

The situation might be further complicated by the presence of other females. Suppose, for example, that the focal female encounters a host that already has eggs in it. Many parasitoids mark hosts after oviposition with a pheromone, so that the second female will be able to ascertain whether or not other eggs are present. If she lays an egg, she is said to **superparasitize** the host. For many years, it was thought that superparasitism was an error in terms of reproductive success. The logic went something like this: if females mark hosts after oviposition, then they are doing it to prevent future oviposition in that host, so ovipositing in a marked host is a mistake. However, Mangel and Roitberg (1988) showed that superparasitism can be adaptive in the sense that a female who superparasitizes may achieve higher fitness than one who does not. We want to be able to predict when a female will superparasitize.

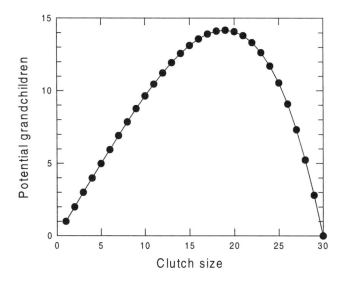

Figure 4.2 The analog of fig. 4.1, using the hypothetical fitness increment given in eq. 4.3.

Finally, some parasitoids make multiple uses of hosts. Upon encounter, a female may lay an egg in a host or feed on that host. The physiological effects of host feeding are manifold. Host-feeding may increase longevity of the parasitoid, may increase egg number, or may affect both. Consequently, we want to be able to predict when a female will feed rather than oviposit in a particular host.

In summary, we want to use dynamic state variable models to determine the ecological and physiological factors shaping (1) how many eggs are laid in a host (how often parasitoids are **egg limited** vs. **time limited**); (2) when an individual superparasitizes; and (3) when an individual host-feeds (Heimpel and Rosenheim 1995). To answer these questions, we will develop a sequence of models of increasing complexity. As described in chapter 3, we use the models to make qualitative predictions about the results of experimental manipulations; then these will be compared with the experimental results.

In the course of developing some of these models, it will be instructive to have an example with a larger range of potential clutches. Thus, in addition to $f(c) = cE(S(c))$ defined by eqs. 4.1 and 4.2, we will work with a "hypothetical parasitoid" for which the fitness increment is (fig. 4.2)

$$f_h(c) = c\left[1 - \left(\frac{c}{30}\right)^3\right] \qquad (4.3)$$

The choices of 30 and the exponent 3 in eq. 4.3 are arbitrary (we picked them to give a range of clutches and a lack of symmetry). In this case, single-host clutches of 19 are optimal. It is not uncommon for certain parasitoids to lay

Table 4.1 Single host optimum clutches for superparasitism

Data of Rosenheim and Rosen			Hypothetical parasitoid		
		Optimal			Optimal
c_0	clutch	Fitness	c_0	clutch	Fitness
0^a	3	29.1	0^a	19	14.2
1	2	19.5	5	15	10.6
2	2	11.4	10	12	7.4
3	1	5.7	15	8	4.4
4^b	1	1.6	20	5	2.1
			25	8	0.7

a. Setting $c_0 = 0$ is a check on previous calculations (always a good thing to do).
b. This is the maximum number of other eggs that allow any fitness to the focal female.

clutches of this size (e.g., Klomp and Teernik 1967; Bai et al. 1992; Vet et al. 1993).

To conclude this section, we note that our assumption that the fitness increment from a clutch can be specified as the potential number of resulting granddaughters is not entirely consistent with the usual notion of fitness in life-history theory. To achieve consistency, we would also have to consider future generations. The method described in section 12.3 could be used for this purpose, although this would require additional information about the biology of *Aphytis*. We do not expect that this would change the qualitative predictions of the models, but this needs to be checked in future work.

4.2 Fixed clutch models

We have already described the simplest fixed-clutch model, which is the single-host maximum clutch. This model predicts that a female *Aphytis* will lay three eggs in a host that does not have any other eggs in it. If we assume that when a female encounters a previously parasitized host, she can sense the number of other eggs in it, then the fitness that she obtains from laying a clutch of size c in a host that already has c_0 eggs in it is given by

$$f_{\text{sup}}(c) = cE(S(c + c_0)) \tag{4.4}$$

In this case, we predict that the female will lay either one or two eggs, depending upon how many eggs are already in the host (table 4.1)

An analogous calculation can be done using the fitness increment for the hypothetical parasitoid, in which case we write

$$f_{\text{sup},h}(c) = c\left[1 - \left(\frac{c + c_0}{30}\right)^3\right] \tag{4.5}$$

The SHM model predicts that clutches will be fixed, independent of phys-iological variables such as the parasitoid's age and egg complement and in-dependent of ecological variables such as time of season, encounter rate with hosts, or mortality rate.

A model that involves at least one ecological variable is the **rate-maximizing** (RM) model. The notion here, lifted from classical diet choice theory (see Stephens and Krebs 1986 or the appendix to chapter 1 here), is that natural selection acts on the rate of accumulation of fitness, so that we predict in-dividuals will behave in a manner that maximizes that rate. We focus only on hosts that are previously unparasitized. The rate of gain of fitness from oviposition of a clutch of size c is given by

$$R(c) = \frac{f(c)}{\text{search time} + \text{handling time}} \tag{4.6}$$

where the search time is the time needed to find a host and the handling time is the amount of time needed to lay a clutch of size c. Rosenheim and Rosen discovered that the first egg in a clutch took 6.5 minutes and each subsequent egg took 3.25 minutes. Thus, we define one period of time as 3.25 minutes; the handling time $h(c)$ associated with a clutch of size c is given by

$$h(c) = 2 + (c - 1) = c + 1 \tag{4.7}$$

Instead of search time, we use encounter rate ρ, which has units of hosts/time period. Thus

$$\text{Search time} = \frac{1}{\rho} \tag{4.8}$$

and eq. 4.6 becomes

$$R(c) = \frac{f(c)}{\frac{1}{\rho} + h(c)} = \frac{f(c)\rho}{1 + \rho h(c)} \tag{4.9}$$

For the hypothetical parasitoid, we assumed that a clutch of size c requires $0.1(c + 1)$ time units. The results (fig. 4.3) are instructive. First, consider *Aphytis*, for which the rate-maximizing clutch is two eggs over a very wide range of encounter rates (fig. 4.3a). In fact, it is only when encounter rates are very low (less than 0.08) that the rate-maximizing clutch shifts to three eggs, which is also the single-host maximum clutch. Thus, we predict that clutch size will decrease as the encounter rate increases, but this might be difficult to verify with *Aphytis* because (1) the shift will only be from three to two eggs and (2) to observe the shift at all, one might need to use encounter rates that are so low that they are unnatural.

The results are more dramatic with the hypothetical parasitoid (fig. 4.3b), for which the SHM clutch is 19 eggs. In this case, the clutch size drops to 10 eggs as the encounter rate increases.

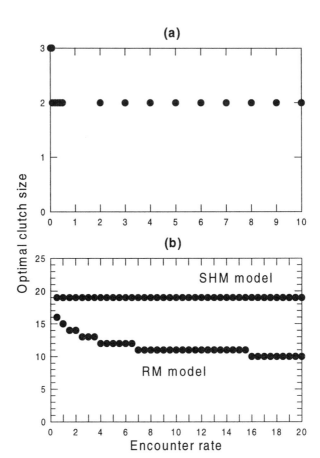

Figure 4.3 The single host maximum (SHM) and rate-maximizing (RM) models can be separated by different qualitative predictions relating encounter rate and clutch size. (a) For the *Aphytis* data, over most of the range, the RM clutch is two eggs and only approaches the SHM clutch for very low encounter rates. (b) The separation of SHM and RM models is more dramatic with the hypothetical parasitoid.

In summary, then, we conclude that SHM models are not sensitive to the encounter rate whereas rate-maximizing models are. For both models, super-parasitism may occur (see below) but will be independent of time, and clutch size will not depend upon time, mortality rate, or egg complement. Wilson and Lessells (1994) and Wilson (1994) refer to both of these as examples of **static optimality models**.

Wilson (1994) tested these ideas (and many more) using the bean beetle *Callosobruchus maculatus*. Females lay their eggs on bean seeds, and the larvae burrow into the seeds on which they then feed. Pupation takes about 25 days; reproductively mature adults weigh 2–10 mg and live for 7–10 days. Females have about 80 eggs, and the clutch size varies between and 1 and

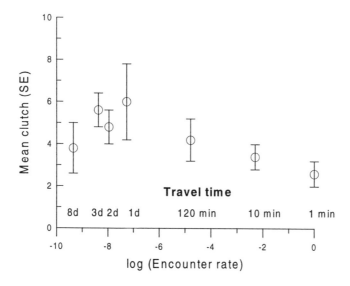

Figure 4.4 Wilson (1994) measured how clutch size depended upon encounter rate for a bean beetle. Consistent with the rate-maximizing model but not with the single host maximum model, clutch size decreased as encounter rate increased.

10 eggs per bean. Wilson found that larval competition leads to fitness increments similar to figs. 4.1 or 4.2 (Wilson 1994, fig. 4).

Wilson studied the effect of encounter rate on clutch size by mating virgin females and allowing them to lay a clutch two hours after mating. The clutch was considered complete when a female walked at least one seed length away from the bean seed. Subsequent seeds were presented to females at 1, 10, 120, or 1440 minutes after the first seed. Wilson estimated oviposition time by using the average time per seed for the first five clutches laid. First clutches did not vary with the travel time manipulation, but subsequent clutches did (fig. 4.4), qualitatively consistent with the rate-maximizing model but inconsistent with the single host maximum model. To quote Clark, "I should hope so!" Other tests of these ideas are found in Nakamura (1997) and Visser and Rosenheim (1998).

4.3 A dynamic but state-independent model

Now we turn to a dynamic but state-independent model. This allows us to consider effects of the time interval and mortality rates but not egg complement. We do this using a model involving superparasitism in a parasitoid that only lays one egg per host. In such a case, let f_u denote the average increment in fitness from oviposition in an unparasitized host. It is often true among such solitary parasitoids that only one egg emerges from a host (the offspring have a contest within the host). Thus, if a female lays an egg in a previously parasitized host, on average she will receive a smaller fitness increment than

f_u. We let f_p denote the average fitness increment from oviposition in a previously parasitized host. Both hosts require the same amount of time τ for oviposition. Finally, we assume that hosts are encountered singly; this makes the analysis easier.

Now there are two encounter rates, ρ_u and ρ_p, with unparasitized or previously parasitized hosts, respectively. Analogous to the classic rate-maximizing solution for the two-prey diet choice problem (Stephens and Krebs 1986), we focus on the rate of gain of fitness R_u if only unparasitized hosts are attacked, and on the rate of gain R_b, if both unparasitized and previously parasitized hosts are attacked.

We find R_b in the following manner (see the appendix to chapter 1). The time interval of length T can be broken into search time S and handling time H, so that $S + H = T$. Given that the search time is S, the parasitoid encounters $\rho_u S$ unparasitized hosts and $\rho_p S$ previously parasitized hosts. Since each host requires τ time units for handling, the handling time is given by

$$H = \rho_u S\tau + \rho_p S\tau = S\tau(\rho_u + \rho_p) \tag{4.10}$$

Assuming that the only activities are search and handling

$$T = S + H = S + S\tau(\rho_u + \rho_p) \tag{4.11}$$

so that

$$S = \frac{T}{1 + \tau(\rho_u + \rho_p)} \tag{4.12}$$

The gain in fitness from unparasitized hosts is $f_u\rho_u S$ and from previously parasitized hosts is $f_p\rho_p S$. Consequently the average fitness acquired over the interval 0 to T is given by

$$(f_u\rho_u + f_p\rho_p)S = (f_u\rho_u + f_p\rho_p)\frac{T}{1 + \tau(\rho_u + \rho_p)} \tag{4.13}$$

and the rate of gain of fitness is given by

$$R_b = \frac{f_u\rho_u + f_p\rho_p}{1 + \tau(\rho_u + \rho_p)} \tag{4.14}$$

Repeating this analysis shows that

$$R_u = \frac{f_u\rho_u}{1 + \tau\rho_u} \tag{4.15}$$

We predict that the parasitoid will superparasitize if $R_b > R_u$. This condition is the same as

$$\rho_u < \frac{1}{\tau}\frac{f_p}{f_u - f_p} \tag{4.16}$$

The right-hand side of eq. 4.16 is a switching value of ρ_u: if ρ_u exceeds the switching value, then it is predicted that the parasitoid will avoid superparasitizing; if ρ_u is less than the switching value, it is predicted that the parasitoid will superparasitize.

In summary, the rate-maximizing model predicts that (1) the parasitoid will always attack all unparasitized hosts; (2) if the encounter rate with unparasitized hosts is sufficiently low, the parasitoid will also attack previously parasitized hosts; (3) the encounter rate with previously parasitized hosts has no effect on the acceptance or rejection of those hosts; and (4) neither mortality nor time within the season will affect oviposition behavior. These predictions are analogous to those obtained in the theory of diet choice (Stephens and Krebs 1986). Now we develop a simple dynamic model for this situation and ignore physiological state but take time into account. Two additional variables are needed. First, we must characterize mortality. We assume that the rate of mortality while searching is m, in the sense that

$$\Pr\{\text{parasitoid survives one time unit}\} = e^{-m} \qquad (4.17)$$

Similarly, we assume that mortality during oviposition is m_{ov} and survival is determined by an expression similar to eq. 4.17. Mortality is included in the rate-maximizing solution in a simple way: the expected lifetime of the parasitoid is $1/m$, so that the lifetime fitness for the rate-maximizing solution is $\frac{1}{m}R_u$ or $\frac{1}{m}R_b$, and hence m has no effect on the optimal behavior. For further discussion, see Mangel (1989).

Because the dynamic model uses discrete time rather than the continuous time implicit in the rate-maximizing solution, we need to characterize the probability of encountering a host in one time unit. Assuming that search is random, we set

$$\lambda_u = \Pr\{\text{parasitoid encounters an unparasitized host in one}$$
$$\text{time unit of search}\} = (1 - e^{-\rho_u - \rho_p})\frac{\rho_u}{\rho_u + \rho_p}$$

$$\tag{4.18}$$

$$\lambda_p = \Pr\{\text{parasitoid encounters a previously parasitized host}$$
$$\text{in one time unit of search}\} = (1 - e^{-\rho_u - \rho_p})\frac{\rho_p}{\rho_u + \rho_p}$$

The logic behind eq. 4.18 is that the chance of encountering a host of either type in one time unit is $1 - e^{-\rho_u - \rho_p}$. Given that a host is encountered, the chance that it is unparasitized is $\frac{\rho_u}{\rho_u + \rho_p}$, and the chance that it is parasitized is $\frac{\rho_p}{\rho_u + \rho_p}$. Recall that when x is small, $e^{-x} \approx 1 - x$, so that when ρ_u and ρ_p are small, the encounter probabilities are approximately ρ_u and ρ_p, respectively.

The fitness measure is

$$F(t) = \text{expected accumulated fitness from oviposition from } t \text{ to } T \quad (4.19)$$

We assume that no oviposition occurs at time T, so that $F(T) = 0$. For previous times, it makes sense to assume that when an unparasitized host is encountered, it is always attacked, but that when a previously parasitized host is encountered, oviposition is chosen to maximize expected reproductive success. With these assumptions, the dynamic programming equation is

$$F(t) = (1 - \lambda_u - \lambda_p)e^{-m}F(t+1) + \lambda_u\{f_u + e^{-m_{\mathrm{ov}}}F(t+\tau)\}$$

$$+ \lambda_p \max\{f_p + e^{-m_{\mathrm{ov}}}F(t+\tau), e^{-m}F(t+1)\} \quad (4.20)$$

The three terms on the right-hand side of eq. 4.20 correspond to not encountering a host, encountering an unparasitized host, or encountering a previously parasitized host during period t. When an unparasitized host is encountered, the parasitoid receives an immediate increment in current fitness f_u and future fitness $e^{-m_{\mathrm{ov}}}F(t+\tau)$, taking into account survival. When a previously parasitized host is encountered, oviposition behavior involves the trade-off between current and future fitness.

The solution of eq. 4.20 generates a value t^* before which previously parasitized hosts will not be attacked and after which they will be attacked. The boundary depends upon the fitness increment from the two kinds of hosts, the encounter rate with each kind, and the two mortality rates. We focus on time and mortality while searching. In fig. 4.5a, we show the boundary between superparasitizing and not superparasitizing for $f_u = 1$, $f_p = 0.2$, $t = 1$, $T = 30$, $m_{\mathrm{ov}} = 0.23$, and $\rho_u = \rho_p = 0.5$ (for which $\lambda_u = \lambda_p = 0.316$). The switching value for the rate-maximizing model is $\rho_u = 0.25$, so based on the rate-maximizing theory, we predict that the parasitoids will never superparasitize. This state-independent dynamic model, on the other hand, predicts that if time is short—that is, if t is close enough to T, the parasitoid will superparasitize. Similarly, if mortality while searching is sufficiently high, it is predicted that the parasitoid will superparasitize regardless of the value of t.

The precise location of the boundary depends upon the encounter rates. For example, we might envision two scenarios about encounter rates. In the "Good world" scenario, $\rho_u = 1.5$ and $\rho_p = 0.5$, whereas in the "Bad world" scenario, $\rho_u = 0.5$ and $\rho_p = 1.5$. The boundary between superparasitizing or not has the same shape as before (fig. 4.5b), but the location depends upon the encounter rates. In the "Good world" scenario, parasitoids wait longer or must experience a higher rate of mortality, while searching, than in the "Bad world" scenario.

Thus, a simple dynamic but state-independent model predicts that superparasitism behavior is dynamic and will respond to time within the season and mortality during search. The rate-maximizing model predicts fixed behavior, regardless of time within the season or mortality during search. Roitberg et al. (1992, 1993) and Fletcher et al. (1994) tested these ideas.

Roitberg et al. (1992) used the solitary drosophilid parasitoid *Leptopilina*

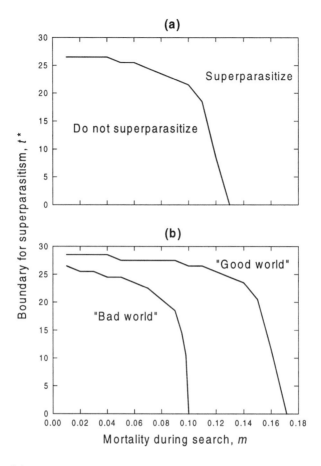

Figure 4.5 (a) The time at which the parasitoid starts accepting previously parasitized hosts depends upon the ecological and physiological factors. Here we hold all but one constant ($f_u = 1$, $f_p = 0.2$, $t = 1$, $T = 30$, $m_{ov} = 0.23$, and $\rho_u = \rho_p = 0.5$) and vary the mortality rate while searching. (b) The location of the boundary depends upon encounter rates. In the "Good world" scenario, $\rho_u = 1.5$ and $\rho_p = 0.5$, whereas in the "Bad world" scenario, $\rho_u = 0.5$ and $\rho_p = 1.5$.

heterotoma, whose larvae are solitary. This wasp has about 800 eggs upon emergence; consequently it is appropriate to assume that the effects of physiological state (egg complement) can be ignored. Females seek out host patches (rotting mushrooms) and search for hosts (*Drosophilid* larvae) by inserting their ovipositors into the patches that may contain larvae. Upon contact with larvae, a female paralyzes and parasitizes the host. Because of the venom in a previously attacked host, a female can recognize a previously parasitized host.

A Dutch (i.e., temperate zone) strain of *Leptopilina heterotoma* was reared on larvae of *D. simulans* under two sets of light to dark cycles: 16h : 8h ("Summer") and 12h : 12h ("Fall"). Since many species of insects are sensitive to photoperiod, Roitberg et al. (1992) assumed that the fall photoperiod indicates that t is approaching T, whereas the summer photoperiod indicates that t is far from T.

The protocol involved a three-day experimental period. On the first two days of the experiment, 4- and 5-day-old wasps were individually placed on a yeast patch containing thirty 48-h-old *D. simulans* larvae. Wasps reared on the 16:8 light cycle were further divided into "Good world" and "Bad world" groups. In the "Good world" treatment wasps were released on yeast patches that contained 30 unparasitized hosts, whereas wasps in "Bad world" treatments were released on patches that contained 30 already parasitized hosts. Wasps reared on the 12:12 light cycle experienced only "Good world" conditions. On the third day, all wasps were individually placed on patches containing thirty *D. simulans* larvae that had already been parasitized by other *L. heterotoma* females. Residence times and superparasitizations were observed and recorded for each wasp.

In a second set of experiments, Roiterg et al. (1992) manipulated perceived mortality by raising parasitoids under both "Good world" and "Bad world" conditions and under steady barometric pressure ("Steady"), typical of a fair summer day, or dropping barometric pressure ("Dropping"), as would occur several hours before the onset of a storm. Such summertime storms are known to be a source of mortality for small insects (Wellington 1946).

The results (fig. 4.6) are striking, and in accord with the qualitative predictions of the dynamic, state-independent model. First (fig. 4.6a), with "Summer" photoperiods, encounter rates clearly affect superparasitisms ($p < .05$ for "Summer, Good world" vs. "Summer, Bad world"). Second, when encounter rates are the same, the closeness of t to T clearly affects superparasitisms ($p < .005$ for "Summer, Good world" vs. "Fall, Good world"). Third (fig. 4.6b), when a cue indicates an imminent increase in mortality, the number of superparasitisms increases ($p < .0014$ for "Steady" vs. "Dropping"). Further intuition about these patterns can be developed by considering the third term in eq. 4.20. This term involves a trade-off between current and future reproduction. Whenever empirical manipulations make future reproduction less valuable (e.g., by increasing m or t), the balance shifts toward current reproduction and superparasitism. Fletcher et al. (1994) investigated this notion by varying the food supply of the parasitic wasp *Venturia canescens*. Some wasps were fed a 50% honey solution, whereas others were fed only water. The starved wasps had a maximum life span of three days and an average life span of about two days, whereas the fed wasps had a maximum life span of five days and an average life span of about four days ($p < .0001$). Fed and starved parasitoids did not differ in egg complement (about 50 eggs each; $p = .65$). Oviposition behavior upon encounter with hosts in a system similar to that used by Roitberg et al. was observed. The starved wasps superparasitized approximately 63% of the hosts they encountered, whereas the fed wasps superparasitized only about 43% of the hosts they encountered and this difference was highly statistically significant ($p < .001$).

Thus we conclude that oviposition behavior is more effectively described as

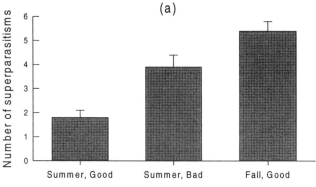

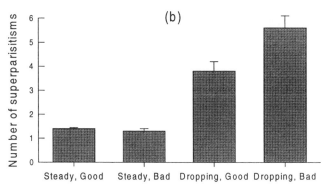

Figure 4.6 The results of experiments by Roitberg et al. (1992, 1993) clearly demonstrate that (i) encounter rates affect the tendency to superparasitize (panel a, Summer, Good world versus Summer, Bad world); (ii) when encounter rates are the same, the closeness of t to T affects the tendency to superparasitize (panel a, Summer, Good world vs. Fall, Good world); and (iii) mortality rate clearly affects the tendency to superparasitize (panel b).

a dynamic process that responds to the mortality rate and the time of the season than as a static, rate-maximizing process. Next, we consider the effect of physiological state, using a dynamic state variable model.

4.4 The proovigenic parasitoid

For parasitoids that lay clutches the simplest physiological state is the number of eggs that a parasitoid holds (i.e., the egg complement). It is clear that none of the single host maximum, rate-maximizing, or state-independent dynamic models will lead to predictions that oviposition behavior depends upon egg complement because these models simply do not include the egg state. Next we develop a model that does and then describe experiments that tested the major predictions of such a model.

We let

$$X(t) = \text{number of eggs at the start of period } t \tag{4.21}$$

The parasitoid has a maximum number of eggs that it can physically contain; we denote this by x_{max}. The minimum value of $X(t)$ is zero. Parasitoids can be broadly classified into those that emerge with their entire egg complement (**proovigenic**) and those that mature eggs during their lifetimes (**synovigenic**). In this section, we focus on the former, in which case the dynamics of the physiological state are given by

$$X(t + 1) = X(t) - \text{clutch laid in period } t \tag{4.22}$$

Fitness, defined in terms of expected potential number of grandchildren, now depends upon time and state, with the definition

$$F(x, t) = \text{maximum expected accumulated number of potential} \tag{4.23}$$
$$\text{grandoffspring from period } t \text{ to } T, \text{ given that } X(t) = x$$

For simplicity, we consider the situation in which only one host type is encountered and for which the increment in fitness from a clutch of size c is $f(c)$. We assume that laying a clutch of size c requires time $h(c)$, as described by eq. 4.7 (or some appropriate modification). If λ is the probability of encountering a host in one unit of search time and m is mortality rate during search and oviposition, we can immediately write the equation that $F(x, t)$ must satisfy:

$$F(x, t) = (1 - \lambda)e^{-m} F(x, t + 1)$$
$$+ \lambda \max_{c}\{f(c) + e^{-mh(c)} F(x - c, t + h(c))\} \tag{4.24}$$

The first term on the right-hand side of eq. 4.24 corresponds to the case in which no host is encountered. In that situation, if the parasitoid survives period t, she begins period $t + 1$ with the same number of eggs. The second term corresponds to the case in which a host is encountered. In that case, the parasitoid may trade current reproduction (larger clutches) with expected future reproduction (survival times fitness at the end of the current clutch). The solution of eq. 4.24 generates $F(x, t)$ and the optimal clutch $c^*(x, t)$ for each egg complement and time.

In fig. 4.7, we show how the first-period clutch $c^*(x, 1)$ depends upon egg complement and mortality when $\lambda = 0.5$, using the fitness increments for *Aphytis* (panel a) or for the hypothetical parasitoid (panel b). In either case, we predict a shift toward higher clutches as egg complement increases. This

(a)

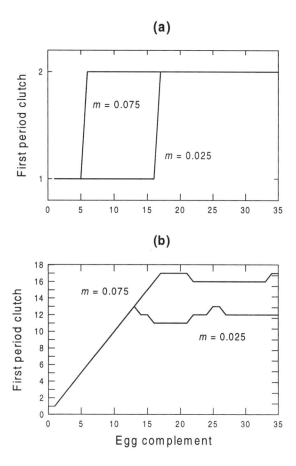

(b)

Figure 4.7 The simplest state variable model predicts that clutch size will respond to egg complement, mortality rate and encounter rate. We show the results for fitness increments associated with *Aphytis* (a) and with the hypothetical parasitoid (b) for $\lambda = 0.5$, $x_{\max} = 35$, and $T = 60$.

shift will occur sooner when mortality rates are higher or (not shown) when encounter rates are lower.

Rosenheim and Rosen (1991) tested the ideas of encounter rate and egg complement using the parasitoid *Aphytis lingnanensis*. This parasitoid is actually synovigenic and may host-feed (see the next sections), so we shall describe the protocols that Rosenheim and Rosen used to ensure that the experimental manipulations were appropriate.

The hosts used in the experiments were large virgin third instar females of the California red scale *Aonidiella aurantii*. Scales were maintained as virgins because the females develop a hard protective shell after mating.

Rosenheim and Rosen provided parasitoids with a uniform history of host encounter but with different egg complements. They manipulated egg load

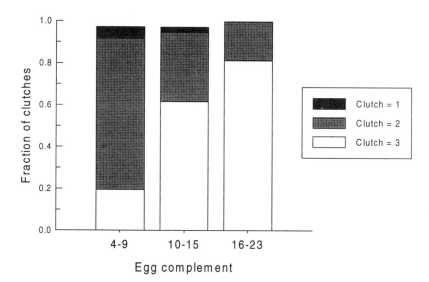

Figure 4.8 Results of the clutch size experiment conducted by Rosenheim and Rosen; details are in the text. We aggregated the egg complement and show the fraction of observations that correspond to clutches of size 1, 2, or 3. For egg complement 4–9, these values are 0.056, 0.722, and 0.194; respectively; for egg complement 10–15 they are 0.025, 0.333, and 0.615 respectively; for egg complement 16–23 they are 0.0, 0.185, and 0.812 respectively. There was one clutch of size 4 for the smaller egg complements.

by using parasitoids of different sizes (see eqs. 4.1, 4.2) or raising parasitoids at low temperature, which slows egg production.

Parasitoids were confined with a single host, and oviposition behavior was observed. After that, the parasitoid was given a second host and only parasitoids ovipositing in both hosts were used in the data collection; after the second oviposition, parasitoids were dissected to determine the egg complement. The egg complement at the start of the experiment was the sum of the eggs laid plus those counted during dissection.

We present results by aggregating the egg complement (fig. 4.8), but the raw data can be found in Hilborn and Mangel (1997, chapter 6) and other versions of the aggregated data in the paper by Rosenheim and Rosen (1991). Virtually all of the clutches were one, two, or three eggs and as egg complement increased, the likelihood of larger clutches increased. Indeed, no clutches of size 1 were observed for an individual that had more than 13 eggs.

In summary, the experiments of Rosenheim and Rosen support the conclusion that oviposition behavior is fundamentally dynamic and responds to changes in physiological state and ecological conditions (encounter rates with hosts and mortality rates). Indeed, egg complement has been suggested as a major source of variability in insect foraging and oviposition behavior (Minkenberg et al. 1992).

Now we can combine the effects of egg load and previous parasitism to extend the results in the last section by assuming that a previously parasitized host already has c_0 eggs in it. The combination of eqs. 4.20 and 4.24 that we choose is

$$F(x, t) = (1 - \lambda_u - \lambda_p)e^{-m}F(x, t + 1)$$

$$+ \lambda_u \max_{c_u}\{f(c_u) + e^{-mh(c_u)}F(x - c_u, t + h(c_u))\} \qquad (4.25)$$

$$+ \lambda_p \max_{c_p}\{f(c_0 + c_p)\frac{c_p}{c_p + c_0} + e^{-mh(c_p)}F(x - c_p, t + h(c_p))\}$$

As before, the first term on the right-hand side corresponds to the situation in which no host is encountered in period t. The second term corresponds to the encounter with an unparasitized host, in which case the clutch laid c_u is determined by the balance between current reproduction $f(c_u)$ and future reproduction $e^{-mh(c_u)}F(x - c_u, t + h(c_u))$. When a previously parasitized host is encountered and the clutch is c_p, the current reproduction is only a fraction of what it would be were the parasitoid only putting her eggs into the host. Hence the term $\frac{c_p}{c_p+c_0}$. Thus, we assume that all offspring survive in a host with $c_p + c_0$ eggs, but the mother is credited only with her share of the associated fitness.

The solution of eq. 4.25 generates a threshold level of eggs $x_p(t)$ required for superparasitism at time t. We applied eq. 4.25 to the fitness increment for *Aphytis* with $c_0 = 1$ and considered "Good worlds" (for which $\lambda_u = 0.8$ and $\lambda_p = 0.2$) or "Bad worlds" (for which $\lambda_u = 0.2$ and $\lambda_p = 0.8$) and two values of the mortality rate (fig. 4.9).

To our knowledge, the combination of experiments of Rosenheim and Rosen and Roitberg et al. that would be needed to demonstrate the existence of the boundary in fig. 4.9 has not yet been done. In chapter 1, we discussed similar boundaries in the feeding behavior of small fish, including experiments that are consistent with the notion of a dynamic threshold, as in fig. 4.9.

There are at least some gamelike aspects to the problem of superparasitism, that is, in eq. 4.25 we assumed that other individuals do not superparasitize (hence the use of $f(c_u)$ for the fitness increment upon encounter with an unparasitized host) and have fixed c_0. These are likely to depend upon the behavior of other individuals, and this would imply that a game is involved (see chapter 10).

4.5 The synovigenic parasitoid: Eggs and reserves

Now we turn to models that involve various complications; these models require more than one state variable. For example, in addition to looking for hosts, parasitoids may look for non-host food sources (e.g., plants that provide nectar or pollen or aphid honeydew). Consumption of non-host foods (or

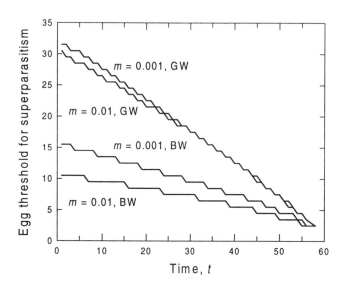

Figure 4.9 A dynamic state variable model for superparasitism generates a boundary in egg complement-time space. At a given time, if the parasitoid's egg complement is above the boundary, we predict that it will superparasitize; if its egg complement is below the boundary, we predict that it will not superparasitize. Other parameters are as in fig. 4.7.

their substitutes in the laboratory) can increase fecundity, longevity, or both (Jervis et al. 1993 and references therein; Rivero-Lynch and Godfray 1997). Alternatively, upon encounter with a host, a parasitoid may feed rather than oviposit (de Bach 1943—a classic paper; Jervis and Kidd 1986 and references therein).

To be explicit, we focus on the case in which reserves are used to "power" the parasitoid in the sense that one period of activity decreases reserves by α; in addition, reserves may be used to increase the egg complement. As before we denote the egg complement at the start of period t by $X(t)$ and introduce

$$Y(t) = \text{amount of reserves at the start of period } t \qquad (4.26)$$

We measure reserves in the same units as eggs and assume that reserves cannot exceed the maximum value y_{max} and that if the reserves fall below a critical level y_c, the parasitoid dies. Finally, we assume that when a non-host food source is encountered, food reserves increase by an amount g. Thus, if a non-host food source is encountered in a period in which x_e eggs are made, the dynamics of $Y(t)$, subject to the constraints concerning y_{max} and y_c, are given by

$$Y(t+1) = Y(t) - \alpha - x_e + g \qquad (4.27)$$

We assume that eggs cannot be converted back into reserves.

As we go through the models, it is important to keep in mind that there are many different ways to formulate the behavior and the physiological dynamics. Consider, for example, how one might change the model if reserves can be used to increase longevity.

Non-host-feeding parasitoids

A parasitoid that does not host-feed needs to find reserves elsewhere. Thus at a particular time, two questions arise: How much of the current reserves should be allocated to producing new eggs? and Should the parasitoid search for hosts or non-host sources of reserves? The key trade-off here is the level of reserves. Since food may not be located immediately, there is value in keeping reserves up. However, keeping reserves at a value that is too high may mean that potential reproduction is lost.

For simplicity of presentation, we assume that the parasitoid is solitary, so that upon encountering a host, she lays a single egg and obtains an increment in fitness f_o from oviposition. In any given time period, the parasitoid can search either for a non-host food source or for a host. We let

$$
\begin{aligned}
\lambda_f &= \text{Probability of encountering a non-host} \\
&\quad \text{food source in a single period of search} \\
\lambda_h &= \text{Probability of encountering a host in a} \\
&\quad \text{single period of search}
\end{aligned}
\tag{4.28}
$$

and introduce fitness

$$
\begin{aligned}
F(x,y,t) &= \text{maximum expected fitness accumulated} \\
&\quad \text{from reproduction between } t \text{ and } T, \text{ given} \\
&\quad \text{that } X(t) = x \text{ and } Y(t) = y
\end{aligned}
\tag{4.29}
$$

which satisfies the end condition $F(x,y,T) = 0$ for every value of x and y. In addition, it satisfies the boundary condition $F(x, y_c, t) = 0$ for every x and t (by measuring x and y in units of eggs, which are integers, we avoid the problems of interpolation near the critical value y_c).

We assume that eggs are matured before she seeks either hosts or food. In this case, the value of seeking a host is given by

$$
\begin{aligned}
V_{\text{host}}(x,y,t) = \max_{x_e}[\lambda_h\{f_o + e^{-m}F(x-1+x_e, y-\alpha-x_e, t+1)\} \\
+ (1-\lambda_h)e^{-m}F(x+x_e, y-\alpha-x_e, t+1)]
\end{aligned}
\tag{4.30}
$$

and the value of seeking a non-host food source is expressed by

$$
\begin{aligned}
V_{\text{food}}(x,y,t) = \max_{x_e}[\lambda_f e^{-m}F(x+x_e, y-\alpha-x_e+g, t+1)\} \\
+ (1-\lambda_f)e^{-m}F(x+x_e, y-\alpha-x_e, t+1)]
\end{aligned}
\tag{4.31}
$$

Then fitness and the optimal behavior are determined according to

$$F(x, y, t) = \max\{V_{\text{host}}(x, y, t), V_{\text{food}}(x, y, t)\} \qquad (4.32)$$

The solution of eq. 4.32 leads to a boundary value of reserves that depends upon egg complement and time (fig. 4.10). When $Y(t)$ is below the reserve boundary, we predict that the parasitoid will search for food sources; otherwise she will search for hosts.

The precise form of the boundary is of less interest to us than its existence. The general prediction is that there is separation into states that correspond to seeking hosts and states that correspond to seeking food. Experiments that demonstrate this separation were conducted by Lewis and Tasuku (1990), Wäckers and Swaans (1993), and Wäckers (1994). These workers investigated the way in which parasitoids use food and host odors, but did not explicitly test the predictions that we have derived. Wäckers and Swaans (1993) and Wäckers (1994) separated the parasitoid *Cotesia rubecula* (which attacks the cabbage butterfly *Pieris rapae*) into fed and deprived groups. The fed groups received 70% saccharose solution whereas the deprived groups were fed only water. The sugar is used to "power" the parasitoid, as in our model. Wäckers and Swaans (1993) found that deprived individuals lived an average of 1.6 days. Mated females who were inexperienced with both host and food odors were given a choice of flying toward flowers of rapeseed or flying toward leaves of rapeseed that were damaged by larvae of *P. rapae*. The results (fig. 4.11) are clearly and significantly in accord with the predictions of the theory: about 65% of the deprived parasitoids sought food sources, whereas only 25% of the fed parasitoids did.

Host-feeding parasitoids

Sometimes hosts can be used as food sources, rather than as sites for oviposition. Parasitoids that exploit hosts in this manner are said to be **host-feeding parasitoids**. Some parasitoids actually can use hosts for both oviposition and feeding; see Jervis and Kidd (1986). As in the previous case, there are different ways in which host reserves can be used (Bartlett 1964; Sandlan 1979; Jervis and Kidd 1986) and not all of them will be explored here; this subject has attracted considerable research effort recently. Excellent introductions to the primary literature are Chan (1991); Rosenheim and Rosen (1992); Collier et al. (1994); Heimpel et al. (1994); Collier (1995a,b); Heimpel and Rosenheim (1995); and McGregor 1997; reviews are found in Rosenheim and Heimpel (1994) and Heimpel and Collier (1996).

4.6 Discussion

Since models of host-feeding become complicated very rapidly (see the primary literature cited before), we have no such models in this section. However, we

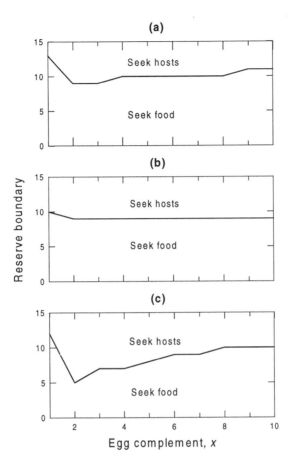

Figure 4.10 The solution of eq. 4.32 generates a boundary in the egg complement-reserve space. Above this boundary, the parasitoid is predicted to seek hosts; below the boundary it is predicted to seek food sources. The results shown used $f_0 = 1.0$, $\alpha = 1.0$, $g = 5.0$, $T = 50$, $y_{max} = 40$, $x_{max} = 10$, and $m = 0.01$, and we show the boundary at $t = 15$ for (a) $\lambda_h = 0.25$, $\lambda_f = 0.25$; (b) $\lambda_h = 0.75$, $\lambda_f = 0.25$; and (c) $\lambda_h = 0.5$, $\lambda_f = 0.5$.

use the intuition developed throughout the chapter to make predictions about the nature of host-feeding and then describe various experiments that have been conducted. We'll see that there is, in fact, incomplete resolution of the question of predictions about host-feeding. Host-feeding implicitly involves two states (egg complement and reserves used for making eggs), and time delays as reserves are converted into eggs. Thus, single host maximum, rate-maximizing, and one-state dynamic state variable models cannot be used to describe host-feeding. At the minimum, a two-state dynamic model is needed.

One of the difficulties in the study of host-feeding is that although egg complement can be determined (it is very tedious), the determination of reserves is still essentially impossible. Thus, one of the state variables is not observable.

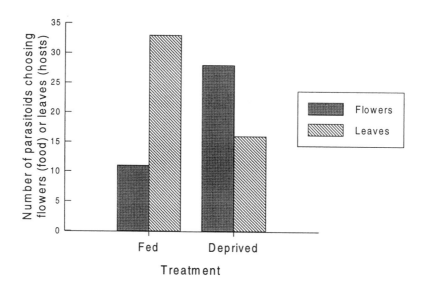

Figure 4.11 Results of the experiments of Wäckers (1994) in which parasitoids were either fed or deprived and then given a choice of seeking hosts or seeking non-host food sources. The deprived parasitoids significantly ($p < .05$, binomial test) sought food sources (flowers) rather than hosts (leaves).

However, we can begin to think about the implications of host-feeding and, based on the experience in this chapter, derive certain qualitative predictions.

Reserves are ultimately converted into eggs for future oviposition. Collier (1995b) showed that when *Aphytis melinus* (a parasitoid of California red scale *Aonidiella aurantii*) host-fed, the host meal led to two new eggs about 15 hours later. These parasitoids can live for 40 days (Collier 1995b) and encounter hosts in the field at the rate of one host every two hours or so (Heimpel et al. 1996 found that one host was encountered every 137 minutes). Thus we anticipate that a parasitoid which is far from T may host-feed when the egg complement is low, to obtain reserves that will be used as eggs in future ovipositions.

Prediction #1: Parasitoids will host-feed when egg complements are low.
Collier et al. (1994) developed a number of simple dynamic state variable models that lead to this prediction. Perhaps their most important conclusion is that there is a threshold egg level x^*, above which the parasitoid oviposits and below which it host-feeds. Furthermore, Collier et al. show that if the delay between host-feeding and having the egg available is a single period, then $x^* = 1$. However, if the delay is greater (e.g., it takes a number of physiological "stages" to convert reserves into eggs), then x^* may be greater than one. Collier (1995a) developed more complicated state variable models, which include physiological realism, that give the same qualitative feature.

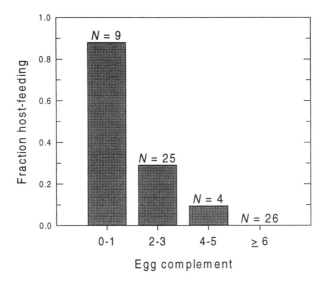

Figure 4.12 The parasitoid *Aphytis melinus* was more likely to host feed at low than at higher values of egg complement (data from Collier et al. 1994). Note, in fact, that parasitoids with more than five eggs never host-fed. The numbers above the bars indicate the number of parasitoids sampled.

Collier et al. (1994) tested these predictions using the parasitoid *Aphytis melinus*. They used newly mated, young females placed into small petri dishes containing scale insect hosts on lemon. They observed parasitoids ovipositing and host-feeding. Once the parasitoid host-fed, it was removed and dissected so that egg complement could be determined. The result (fig. 4.12) shows a clear qualitative agreement with the prediction: as egg complement increased, the likelihood that a parasitoid host-fed decreased. Heimpel and Rosenheim (1995), using *Aphytis melinus* that attacks oleander scale, found results consistent with those of Collier et al.: egg complement significantly ($p < .001$) affected whether a parasitoid host-fed or not.

The results of Collier et al. are in contrast to those of Rosenheim and Rosen (1992) who found that egg complement did not influence (in the sense of a statistically significant result) the host-feeding behavior of the parasitoid *Aphytis lingnanensis*. Collier et al. note the following differences between their work and that of Rosenheim and Rosen. First, Rosenheim and Rosen examined behavior of the parasitoid on the first host encounter whereas Collier et al. waited until the parasitoid host-fed, which usually took two or three encounters. Thus, in the experiments of Collier et al., parasitoids differed in both experience (encounters with hosts) and egg complement. Second, Rosenheim and Rosen used second instar scale insects, whereas Collier et al. used third instar scale insects; the second instar scales are poor quality for oviposition purposes, and Collier et al. suggest that host-feeding on third instar scales—which are high-quality hosts for oviposition purposes—may be

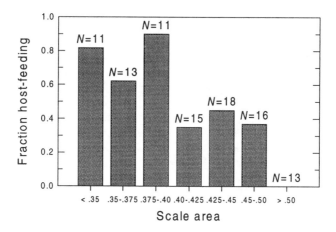

Figure 4.13 Rosenheim and Rosen (1992) found that smaller second instar hosts were used for host feeding with much higher frequency ($p < .001$) than larger hosts. Scale area is measured in mm^2.

sensitive to egg complement, whereas host-feeding on second instar scales may not be. The question has not yet been resolved.

However, the notion that host variation may lead to different behavioral responses is valuable. For example, scale area affects the number of progeny that emerge. Heimpel et al. (1996) found that oviposition in first instar scales leads to a fitness increment (potential grandchildren) of 3 emerging, in second instar scales to a fitness increment of 9 progeny emerging, and in third instar scales to a fitness increment of 11. Thus we come to the second qualitative prediction.

Prediction #2. Smaller hosts are more likely to be used for host-feeding than for oviposition.
Rosenheim and Rosen (1992) found exactly this dependence (fig. 4.13) and used stepwise logistic regression to determine that host area was the most important variable and highly significant ($p < .001$) as a predictor of whether a parasitoid would host-feed. Rosenheim and Rosen used *A. lingnanensis*; Heimpel and Rosenheim (1995) confirmed the result with *A. melinus*.

Into the woods

All of the studies described thus far took place in the laboratory. Heimpel et al. (1996, 1998) combined predictions 1 and 2 and took them to the field. During three seasons, parasitoids were observed in an abandoned almond orchard. Once found, parasitoids were followed until a host encounter, and their behavior was observed using a handlens. After oviposition, parasitoids and the encountered host were captured and brought to the laboratory for dissection. In this manner, the egg complement of the parasitoid at the time of encounter was determined. Host area (treated as an ellipse) was an index of host size. Approximately 70 parasitoids were observed either ovipositing or host-feeding.

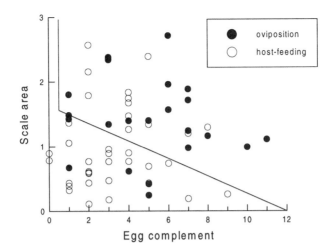

Figure 4.14 Heimpel et al. (1996) studied the oviposition and host-feeding behavior of *Aphytis melinus* in the field and found support for the predictions developed in this chapter. Filled circles denote hosts that were used for oviposition, open circles hosts that were used for host-feeding. The boundary line between oviposition and host-feeding is drawn for ease of presentation and was determined by using least squares to determine the parameters S_o and E_o that give the best fit between the data and the line $S_b(E) = S_o(1 - E/E_o)$, with the understanding it is predicted that the hosts which fall above the line are used for oviposition and hosts below the line are used for host-feeding. Oviposition is impossible when the egg complement is less than one, hence the vertical line.

The broad general prediction is that parasitoids with larger egg loads are more likely to oviposit and that smaller hosts are more likely to be used for host-feeding than oviposition. The field studies support (fig. 4.14) all of the predictions that we've developed in this chapter:

- Ovipositing parasitoids had a higher mean egg load (5.1 ± 0.51, SEM) than host-feeding parasitoids (3.4 ± 0.32).
- Scale hosts used for oviposition were larger (1.5 ± 0.12 mm^2) than scale insects used for host-feeding (1.0 ± 0.11 mm^2).
- Both scale area and egg complement entered into a logistic regression for the probability of host-feeding versus ovipositing, and both were significant ($p = .004$ and $p = .039$, respectively).
- The interaction term (scale area) × (egg complement) did not enter into the logistic regression significantly ($p > .5$), showing that the two variables affect behavior independently.

5

Winter Survival Strategies

Techniques and concepts used in Chapter 5

Day–night coupling (5.1, 5.4)
Sensitivity testing (5.1, 5.8)
Interpolation in simulations (5.1)
Dynamic semigame model (5.2)
Removal experiments (5.3)
Multidimensional state variables (5.4, 5.5)
Allocation of activity times (5.5)
Food deprivation experiments (5.8)

In this chapter we use dynamic state variable models to study winter foraging and fat-reserve strategies for two vertebrate species: willow tits (*Parus montanus*) and Atlantic salmon (*Salmo salar*). In both examples we define fitness in terms of winter survival. Both species are subject to predation risk while foraging, so they both face critical trade-offs between starvation and predation. The main source of stochasticity assumed to face willow tits in winter is variable costs of thermoregulation induced by cold weather periods. For the salmon, daily foraging success is variable. Field or laboratory tests of model predictions are also described.

5.1 Foraging and fattening strategies for willow tits

The willow tit is a small (ca. 11-g) passerine that resides in the northern forests of Eurasia. In appearance it closely resembles the black-capped chickadee (*Poecile atricapillus*), which inhabits similar environments in North America. The behavior of willow tits has been under intensive investigation for several decades, especially in Scandinavia (e.g. Haftorn 1956, 1992; Hogstad 1988; Ekman and Lilliendahl 1993; Brodin 1994). This research provides useful information for formulating and testing state variable models.

First we discuss a simple model of the winter foraging and fattening strategy of an individual willow tit. This model omits important factors such as hoarding and social interactions; these will be discussed later using more complex models. The simple model is useful in helping to understand some of the main trade-offs that affect the evolution of behavior in this species. Extensive theoretical investigation of the foraging behavior of small birds in winter appears in Houston and McNamara (1993); McNamara et al. (1994); Bednekoff and Houston (1994a), and other papers by these authors.

For this initial model, covering a 120-day period in midwinter, we make the following assumptions.

1. Night-time metabolic costs fluctuate randomly as a result of weather conditions but with no seasonal trend. In mathematical terminology, the probability distribution for metabolic costs is stationary—that is, time-independent. Bednekoff and Houston (1994b) describe a model of fattening strategies in nonstationary environments.

2. During daylight hours, the bird can choose between three habitats H_0, H_1, and H_2. H_0 is a refuge that has no food but is safe from predators, H_1 (interior branches of trees) has limited food resources and a relatively low predation risk, and H_3 (exterior branches) has more abundant food but a higher predation risk. Willow tits are subject to heavy predation by avian predators such as pygmy owls (*Glaucidium passerinum*) and hence are presumably at greatest risk when foraging in exposed outer branches.

3. Daily metabolic cost and predation risk are increasing functions of the level of fat reserves (Lima 1986; Houston and McNamara 1993; Bednekoff and Houston 1994a,b; McNamara et al. 1994). Direct empirical evidence of a positive correlation between fat reserves and predation risk is lacking and probably difficult to obtain. Metcalfe and Ure (1995) measured take-off speed and maneuverability of zebra finches (*Taeniopygia guttata*) as functions of body mass. They found that a relatively small increase (about 7%) in body mass had substantial effects on flight performance and concluded that "small birds are inherently more vulnerable at dusk [when their body masses are at a maximum] than at dawn" (ibid, p. 396). Bednekoff (1996) suggested that the implications of reduced flight performance on predation risk would be strong.

4. During the winter the bird's fitness can be characterized as its probability of surviving.

The state variable is $X(t, d)$ = energy reserves at the start of period t on day d; see tables 5.1 and 5.2 for parameters and variables used in the model. Nighttime metabolic costs are stochastic and may equal c_b ("bad night") with probability p_b or c_g ("good night") with probability $p_g = 1 - p_b$. If possible,

Table 5.1 Parameter values for the basic model of willow tit foraging and fattening strategies[a]

Parameter	Meaning	Value
x_{max}	Maximum fat reserves	$2.4\,g$
μ_{10}	Basic daily predation risk, habitat H_1	$.001/d$
μ_{20}	Basic daily predation risk, habitat H_2	$.005/d$
λ	Rate of increase of predation risk with body mass	$0.46\,g^{-1}$
e_1	Net daily forage intake, habitat H_1	$0.6\,g/d$
e_2	Net daily forage intake, habitat H_2	$2.0\,g/d$
c_g	Nighttime metabolic cost, good conditions	$0.48\,g$
c_b	Nighttime metabolic cost, bad conditions	$1.20\,g$
p_b	Frequency of bad conditions	0.167
m_0	Mass of bird with zero fat reserves	$10\,g$
γ	Metabolic rate	$0.04/d$
T	Number of time periods per day	50
D	Number of days in winter	120

a. Clark and Ekman 1995; Brodin and Clark 1997.

Table 5.2 Symbols used in the basic model (see also table 5.1)

Symbol	Meaning
d	Day in winter, $1 \leq d \leq D$
t	Time period during day, $1 \leq t \leq T$
$X(t,d)$	Energy reserves at beginning of period t on day d (g)
$F(x,t,d)$	Fitness, that is, probability of survival from (t,d) to period 1 on day $D+1$, given $X(t,d) = x$
H_i	Foraging habitat i ($i = 0,1,2$)

the bird should end each day with enough reserves to survive the worst night. Thus the model should predict that $X(T,d) \geq c_b$.

Since there are costs but no advantages to having extra reserves, we expect that, in fact, $X(T,d) = c_b$. The question is, what sequence of foraging in habitats H_1 or H_2 or resting in H_0 will maximize fitness? Because of the costs of having high reserves, we would anticipate that the bird should maintain low reserves for as much of the day as possible and concentrate on feeding near dusk to bring $X(T)$ up to c_b. If we adopt the rate-maximizing perspective (appendix to Ch. 1), we would predict that the bird will choose the habitat in which μ_i/e_i is minimized, for example if $\mu_2/e_2 < \mu_1/e_1$, the optimal daily strategy is to rest (H_0) at the beginning of the day and then forage in H_2 for the remainder of the day. For the parameter values listed in table 5.1, however, $\mu_1/e_1 = 0.0017 < \mu_2/e_2 = 0.0025$, so the preferred habitat is H_1. But H_1 does not have a high enough rate of food recovery to make up for fat reserves lost during a cold night. Consequently, H_2 must be used for at least part of the day to replenish reserves after a cold night.

Although this argument is not completely rigorous (for example, we did not consider mass-dependent costs), the computed solution shown below closely resembles the previous description. Other qualitative predictions will be discussed after we have developed the basic model.

The basic model

We need two time variables: d, the day in winter, and t, the time of day. We have $d = 1, 2, \ldots, D$ and $t = 1, 2, \ldots, T$, where D is the length of winter (120 days) and T is the number of time periods in each day. More precisely, days $d = 1, 2, \ldots, D$ are winter days, and $D + 1$ represents the first day of spring; we consider foraging behavior only for winter days. Similarly, periods $t = 1, 2, \ldots, T$ occur during the day (when foraging takes place), and period $T + 1$ corresponds to the beginning of night.

The state variable $X = X(t, d)$ represents fat reserves (g) at the beginning of period t on day d. In each time period t, the bird chooses one of the three habitats H_i. Daily rates of fat increase while foraging in habitat H_i are denoted by e_i, and predation risk per day is given by $\mu_i(x) = \mu_{i0}e^{\lambda x}$ (see table 5.1). In habitat H_0 by assumption $e_0 = 0$ and $\mu_{00} = 0$. Metabolic costs during the day are $\gamma(m_0 + x)$, where x denotes fat reserves (which vary through the day), and m_0 is the bird's mass at zero reserves. Overnight metabolic costs are denoted by c_g or c_b, corresponding to a "good" or "bad" night, respectively. The probability of a bad night is p_b, and the probability of a good night is $p_g = 1 - p_b$.

Fitness $F(x, t, d)$ is defined as the maximum probability of surviving from period t on day d to the morning of day $D + 1$, given that $X(t, d) = x$. Recalling that $T + 1$ is the beginning of night,

$$F(x, T+1, D) = \begin{cases} 0 & \text{if } x < c_g \\ p_g & \text{if } c_g \leq x < c_b \\ 1 & \text{if } x \geq c_b \end{cases} \tag{5.1}$$

Equation 5.1 says that the bird necessarily dies overnight if its reserves are less than c_g, the nighttime requirement on a good night. Also, the bird dies on a bad night if $x < c_b$.

Similarly for $d < D$,

$$F(x, T+1, d) = \begin{cases} 0 & \text{if } x < c_g \\ p_g F(x - c_g, 1, d+1) & \text{if } c_g \leq x < c_b \\ p_g F(x - c_g, 1, d+1) + p_b F(x - c_b, 1, d+1) & \text{if } x \geq c_b \end{cases} \tag{5.2}$$

Equation 5.2 can be expressed more compactly using expectations:

$$F(x, T+1, d) = E_c F(x - c, 1, d+1)$$

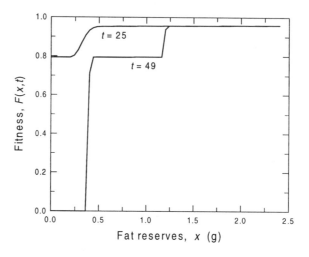

Figure 5.1 The fitness function $F(x, t, d)$ for two values $t = 25$ (midday) and $t = 49$ (just before dark) on day $d = D - 20$. Base-case parameters of the fattening strategy model (table 5.1).

where it is assumed that $F(x - c, 1, d + 1) = 0$ if $x - c < 0$. This form would apply to an arbitrary distribution of nighttime costs c. We will refer to this form again later in eq. 5.6. The dynamic programming equation for $d \leq D$ and $t \leq T$ is

$$F(x, t, d) = \max_i [1 - \delta\mu_i(x)] F(x'_i, t + 1, d) \quad (x \geq 0) \tag{5.3}$$

where

$$x'_i = x + \delta e_i - \gamma\delta(m_0 + x) \text{ with } 0 \leq x'_i \leq x_{\max} \tag{5.4}$$

Here $\delta = 1/T$ is needed to express the daily rates e_i and γ in terms of a single time period t. Note that, for simplicity, here we are assuming deterministic food gain during the day; all stochasticity arises in terms of nighttime costs. At the end of this section, we briefly discuss the possibility of changing this and other model assumptions.

Note that we do not specify $F(0, t, d) = 0$ (as was done in chapter 1), but instead calculate $F(0, t, d)$ using the dynamic programming algorithm. This is an important technical point brought about by the numerical discretization of the continuous state variable x (see section 2.1). Specifically, the values of $F(x'_i, t + 1, d)$ are interpolated in the computer program. If we were to set $F(0, t, d) = 0$, this interpolation would be quite inaccurate since $F(x, t, d)$ does not approach zero, as $x \to 0$ (see fig. 5.1 for $t = 25$). A bird with any reserves at all can avoid starvation by immediately foraging. Hence it is best not to assume that $F(0, t, d) = 0$ but that the bird will starve if $x'_i < 0$— that is, $F(x, t, d) < 0$ for $x < 0$.

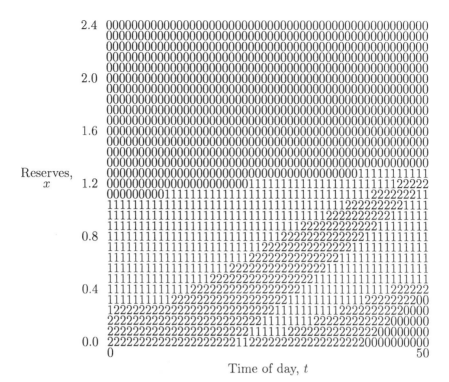

Reserves, x

Time of day, t

Figure 5.2 Decision matrix $H(x, t, d)$ for fattening model (base case). Entries refer to habitat number 0, 1, or 2. (Here $d = D - 20$.)

The same technical point would apply to any model that includes the possibility of starvation. The problem did not arise in the patch-selection model of chapter 1, because only fixed discrete values of $x = 0, 1, 2, \ldots$, and of x' were considered in that example.

Model predictions

The program wtit1.cpp available on the OUP web site was used to generate predictions for the willow-tit model. Figure 5.1 shows the computed fitness function $F(x, t, d)$ for the basic model for $d = D - 20$ and for two values of t, that correspond to midday ($t = T/2$) and just before dusk ($t = T - 1$); we used $T = 50$ periods per day. The curve for $t = 49$ is self-explanatory: unless the bird has built up nearly $c_g = 0.48\,\mathrm{g}$ of reserves by $T - 1$, it can't survive the night (good or bad), and unless it has nearly $c_b = 1.2\,\mathrm{g}$, it can't survive a bad night.

The habitat-decision matrix $H(x, t, d)$ (for $d = D - 20$) is given in fig. 5.2; to understand this matrix, it helps to consider the simulated optimal trajectories of fat reserves shown in fig. 5.3.

First note that these trajectories (which apply to a bird that has just

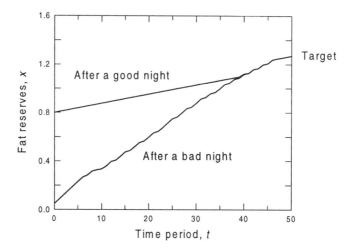

Figure 5.3 Optimal daily trajectories of fat reserves (a) following a good night and (b) following a bad night. Both trajectories achieve the optimal target $X(T) \cong 1.2 \, \text{g}$.

experienced a good or a bad night, respectively) both reach the same "target" $X(T) = 1.21 \, \text{g} \cong c_b$ at the end of the day, as predicted. Figure 5.2 shows that birds that have reserves $x > 1.25 \, \text{g}$ should always rest ($i = 0$). Birds with $x < 1.2 \, \text{g}$ should forage, either in H_1 or H_2. If the current reserve level $X(t, d)$ is sufficiently high, H_1 is preferred (recall that this agrees with the rate-maximization rule). The trajectory labeled "after a good night" in fig. 5.3 uses H_1 for most of the day; the bird is able to reach the target without exposing itself to the dangerous habitat H_2, except near dusk. This luxury is not available after a cold ("bad") night, however. In this case the bird must spend much of the day in H_2 to get enough food.

The optimal trajectory $X(t, d)$ follows right along the boundary between the 1s area and the 2s area shown in fig. 5.2. The bird switches back and forth between H_1 and H_2, building up its reserves at a more-or-less constant rate, to reach the target $X(T) = 1.21 \, \text{g}$. (There is another area of 1s and 2s in fig. 5.2, lying below the upper two areas; these lower areas don't arise in practice but correspond to a desperate bird that tries at least to reach the secondary target $X(T) = 0.48 \, \text{g}$.)

Ekman and Lilliendahl (1993) measured fat-reserve trajectories of willow tits in Sweden; their published results indicate a constant rate of fat buildup during the day, similar to the trajectories shown in fig. 5.3; see fig. 5.5 below.

Interpolating in the forward iterations

In performing the forward iterations shown in fig. 5.3, the computed values of $X(t)$, of course, do not necessarily coincide with any of the discretized values stored in the computer. In this situation it is not clear what decision the bird would use since the decision matrix $D(x, t)$ has been computed only for

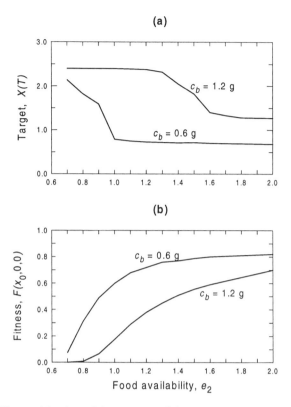

Figure 5.4 Effects of food availability e_2 on (a) target reserves level $X(T)$ and (b) fitness $F(x_0, 1, 1)$. Here $x_0 = 0.5$ g, and $D = 120$ d; two values of c_b ("bad" night cost) are also shown, $c_b = 0.6$ or 1.2 g.

these discrete values. To deal with this, we can use interpolation, similar to the procedure used in the dynamic programming calculation (see chapter 2). Specifically, if the computer value $x_c(t)$ lies between two discretized values j and $j + 1$, we assume that the bird will spend part of period t in each of the two habitats. The proportions of time spent in each habitat are proportional to the interpolation intervals $\Delta x_c = x_c - j$ and $1 - \Delta x_c$, exactly as in the case of linear interpolation. This avoids the necessity of using extremely short time periods. An alternative procedure would be to allow the bird to choose one or the other habitat at random for the whole period with probabilities proportional to the interpolation intervals.

Sensitivity to availability of food

In the base-case model, the feeding-rate parameter $e_2 = 2.0 \, \text{gd}^{-1}$ is large enough to allow the bird to make up the largest overnight deficit $c_b = 1.2$ g in a single day's feeding. As noted above, this implies that the target $X(T)$ equals approximately 1.2 g. Now we ask, how important is the level of food e_2 in terms of model predictions? Figure 5.4(a) shows how the target $X(T)$

depends on e_2 for two cases, $c_b = 1.2$ g (base-case value) and $c_b = 0.6$ g. Note that, for $c_b = 1.2$ g, the target increases sharply when e_2 is below $1.4\,\mathrm{gd}^{-1}$. The explanation is the following. Daytime metabolic costs $\gamma(m_0 + x) = 0.20$–$0.25\,\mathrm{gd}^{-1}$, depending on the level of reserves x. Thus in the case $c = c_b = 1.2$ g, the bird must gain a total of 1.40–$1.45\,\mathrm{gd}^{-1}$ to replenish the high overnight loss. If $e_2 < 1.4$ g, this is impossible: the bird will steadily lose weight in a run of consecutive cold nights. Therefore it pays to carry extra fat reserves as a hedge against this possibility.

The same logic applies to the lower curve in fig. 5.4(a) that corresponds to $c_b = 0.6$ g. Now total daily requirement after cold nights is 0.8–$0.85\,\mathrm{gd}^{-1}$, and the target $X(T)$ again rises sharply if e_2 is less than this.

Food availability e_2 also has a pronounced effect on fitness (i.e., overwinter survival), as shown in fig. 5.4(b). For example, the average daily requirement is $0.833c_g + 0.167c_b + \gamma(m_0 + x) \cong 0.82\,\mathrm{gd}^{-1}$ (when $c_b = 1.2$ g), yet for $e_2 = 0.82\,\mathrm{gd}^{-1}$ the bird's probability of surviving the winter is almost zero. Just being able to meet the average daily winter food requirements is not good enough to ensure survival.

Therefore, sufficiently abundant winter food supplies are extremely important for willow tits. In fact, willow tits often hoard food for winter use. If natural food supplies are barely sufficient for average daily needs, the presence of a relatively small hoard of extra food can make a large difference to winter survival. This will be demonstrated in section 5.3 by using a simple modification of our basic model. Hoarding, however, is a rather complex adaptation. We will also briefly discuss a more complicated model of short-term and long-term hoarding behavior.

Hoarding food is only useful if the hoard can be protected from pilfering. Perhaps for this reason, hoarding parids such as willow tits live in small territorial flocks in winter. One consequence of this social arrangement is discussed in section 5.2.

What do birds die of?

Starvation and predation are the two sources of mortality in our model. The optimal fattening strategy involves a trade-off between these two risks. It might naively be supposed that the optimal combination of risks would occur when the two risks are approximately equal, but this is not the case (McNamara and Houston 1987). In our base-case model, for example, no bird ever dies of starvation! The value $e_2 = 2.0\,\mathrm{gd}^{-1}$ exceeds the sum of daytime and nighttime metabolic costs, even on cold nights. Thus the bird can always gain enough fat reserves each day to survive the night. It faces a predation risk of at most $\mu_2(x) \cong .007\,\mathrm{d}^{-1}$ to obtain these reserves, but this is much lower than the risk of starvation, $p_b = .167$, if it does not carry sufficient reserves $X(T)$. Hence (in the base case) the bird never starves.

In reality, some birds do doubtlessly die of starvation. In section 5.2, for example, we will see that starvation is a real possibility for subordinate willow tits. Starvation becomes possible if food supplies are not large enough to

Table 5.3 Starvation and predation risks over the entire winter, predicted by the willow tit fattening model, as functions of e_2 = foraging rate in H_2. All other parameters are as in the base-case model. (Probabilities are calculated using Monte Carlo forward iteration, $N = 1000$.)

$e_2(\mathrm{gd}^{-1})$	Pr (starvation)	Pr (predation)
2.0	0	.255
1.6	0	.309
1.4	0	.360
1.2	.006	.468
1.0	.077	.582
0.8	.750	.242

replenish reserves in a single day after a severe night. As shown in fig. 5.4, this forces birds to carry extra reserves, but doing so does not guarantee freedom from starvation. Any protracted spell of cold weather can slowly exhaust the bird's reserves and result in starvation. Table 5.3 shows the probabilities of starvation and predation over winter as functions of e_2 = foraging rate in H_2. Unless e_2 is extremely low, birds compensate for limited food supplies by increasing their foraging activity in the dangerous habitat H_2 and thereby incur increased risks of predation. For example, with $e_2 = 1.0\,\mathrm{gd}^{-1}$, overwinter predation risk rises to 58.2%, but the risk of starvation is only 7.7%. When $e_2 = 0.8\,\mathrm{gd}^{-1}$, however, starvation becomes the leading cause of death; in this extreme case, few birds (0.8%) survive the winter, on average.

The preponderance of predation over starvation has an important consequence at the population level. Imagine a local population of willow tits, and let p (Starve) and p (Pred) be the probabilities of overwinter starvation and predation, respectively. In a starvation event, all birds die overwinter, and the local population is wiped out. Predation risk, on the other hand, reduces the population only by the factor p (Pred) each winter. Both probabilities are likely to be greater for juveniles than for adults, as we discuss in the next section. Unless p (Pred) is high, the likelihood of a local population crash is low. It is interesting in this model that individual fitness optimization results in a strong bet-hedging strategy at the population level; see Seger and Brockmann (1987) or Yoshimura and Clark (1993) for discussions of bet hedging.

Sensitivity to predation risk

Gosler et al. (1995) found that Great tits (*Parus major*) in Britain reduced their mean winter fat reserves when the population of sparrow hawks (*Accipiter nisus*) increased following the ban of DDT, which suggests that these birds responded to increased predation risk by foraging less intensively, either by reducing foraging time or by increasing their vigilance while foraging. Parallel observations for willow tits have not been reported, to our knowledge. Our base-case model, however, predicts no response to increased predation risk—birds achieve the optimal end-of-day reserves target $X(T) \cong c_b$ regardless of the level of predation risk (within realistic bounds). As we have seen, this strategy effectively eliminates the risk of starvation, whereas reducing $X(T)$ below c_b would greatly increase the starvation risk.

This model prediction is partly a consequence of our assumption that only two levels of nighttime metabolic cost, c_g and c_b, occur. It would be more realistic to assume that the nighttime cost c has a continuous distribution, for example, normal with mean $\bar{c}$ and variance σ_c^2. The basic model can be modified to allow for this new assumption (a discretized approximation to the normal distribution would be used in the program). Another possible modification would allow for variation in food supplies from day to day, which could be caused by harsh weather conditions such as snow or high winds (see section 5.4).

Such changes in the model would make fitness a more continuous function of fat reserves, which could lead to predictions that would agree qualitatively with the observations of Gosler et al. (1995). Data pertaining to weather variability and the effects of weather on metabolic costs and food availability would be useful. We will not pursue this topic further here, however.

5.2 Social interaction

In winter willow tits live in small, territorial flocks that consist of an adult pair and two to four unrelated juveniles (Ekman 1979; Hogstad 1989). In terms of access to foraging sites, adults dominate juveniles, and males dominate females. Hogstad (1989) observed 78 flocks in central Norway from 1979–1986, and recorded mean winter survival rates of 0.74 and 0.32 for adults and juveniles, respectively (see also Ekman 1990).

Ekman and Lillendahl (1993) measured daily trajectories of body mass in individual willow tits, and found that the body mass of subordinate males regularly exceeded that of dominant males (fig. 5.5). This appears counterintuitive: why should the dominant, whose access to foraging habitats is unlimited, carry lower fat reserves than the subordinate?

The model discussed in section 5.1 can help to understand this phenomenon. With unrestricted access to all foraging habitats H_i, willow tits build up fat reserves to a target level $X(T)$ that ensures overnight survival. Following a cold night, the bird must spend most of the day foraging in H_2 (the richer, but riskier habitat) to replenish its reserves (figs. 5.2 and 5.3). With a dominant–subordinate pair, however, the dominant will monopolize H_2 after a cold night, and the subordinate will be relegated to H_1, which contains insufficient food. Consequently the subordinate needs to hold extra reserves, as a hedge against a sequence of several successive bad nights (see fig. 5.4). Further, a protracted spell of bad nights will eventually exhaust the subordinate's

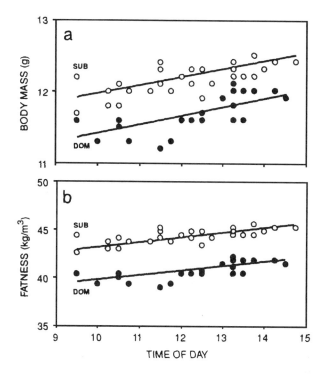

Figure 5.5 Body mass (a) and fatness index (b) for the dominant (filled circles) and subordinate (open circles) for a selected pair of male willow tits. From Ekman and Lilliendahl (1993).

reserves, and this may explain the observed lower survival rate of subordinates. In fact, for the base-case parameters, the subordinate cannot survive more than two consecutive bad nights: the 24-h decrease in reserves X is equal to approximately 0.94 g for a bird restricted to habitat H_1. The probability of three consecutive bad nights at any given time is $p_b^3 = .0046$. However, the probability that three consecutive bad nights will not occur over 120 days is only 62.2%. The subordinate's survival rate must be less than this.

General dynamic state variable games between two players can be quite difficult to analyze (see, for example, Houston and McNamara 1988, who discuss a dynamic version of the famous hawk-dove game). Clark and Ekman (1995) developed a dynamic "semigame" model of dominant and subordinate fattening strategies of willow tits. They assumed that in each time period t the dominant selects its optimal habitat, as in the model of section 5.1. The subordinate is excluded from the dominant's current foraging habitat (H_1 or H_2) but is never excluded from the resting habitat H_0. The subordinate maximizes its fitness, subject to this interference. We call this a semigame because the dominant's behavior is completely independent of the subordinate. (Thus, the model ignores any cost of maintaining dominance.)

We continue to use the notation of the previous section (tables 5.1 and 5.2), and now write $F_{\text{dom}}(x, t, d)$ for the dominant's fitness function and $H_{\text{dom}}(x, t, d)$ for the dominant's habitat-decision matrix. These are the same as $F(x, t, d)$ and $H(x, t, d)$ in section 5.1.

Let $Y(t, d)$ be the subordinate's fat reserves. Then the subordinate's fitness function is $F_{\text{sub}}(y, h, t, d)$ where h is the habitat selected by the dominant in period t, day d. The dynamic programming equations for the subordinate are similar to eqs. 5.1–5.4, except for the restriction of habitat use imposed by the dominant:

$$F_{\text{sub}}(y, h, T+1, D) = \begin{cases} 0 & x < c_g \\ p_g & c_g \leq x < c_b \\ 1 & x \geq c_b \end{cases} \qquad (5.5)$$

$$F_{\text{sub}}(y, h, T+1, d) = E_{c,h'} F_{\text{sub}}(y - c, h', 1, d+1) \qquad (5.6)$$

$$F_{\text{sub}}(y, h, t, d) = \max_{(i)}[1 - \delta\mu_i(x)] E_{h'} F_{\text{sub}}(y_i', h', t+1, d) \ (y \geq 0) \quad (5.7)$$

The notation used here requires some explanation. The right-hand sides of eqs. 5.6 and 5.7 involve h', which denotes the dominant's choice of habitat in the next time period $t + 1$. We do not want to assume that the subordinate knows in advance which habitat the dominant will use next (this would involve knowing the dominant's fat reserves $X(t)$), but we will assume that the subordinate does know the frequencies with which each habitat H_i is used by the dominant. The subordinate can quickly learn this by experience, but in the computer we calculate these frequencies by forward simulation of the dominant's behavior. A smart subordinate might learn the conditional probabilities of the dominant's habitat use, depending on the previous night's temperature, but we do not consider this possibility. In eq. 5.7 the maximization over i (the subordinate's habitat) requires that $i \neq h$ (where h is the dominant's current habitat); this restriction does not apply if $h = 0$.

Model predictions

Figure 5.6 shows the predicted average daily fat reserves for dominant and subordinate birds and the overwinter survival probabilities, as functions of $e_2 = $ fat reserves gained in habitat H_2 (Clark and Ekman 1995; their model used only one decision period per day to simplify the computations). When $e_2 < 1.2$ g, the dominant must maintain high reserves because one day's foraging cannot make up for losses of fat on a cold night. When $e_2 > 1.2$ g, the dominant keeps its reserves close to the target level $X(T) = 1.2$ g. The subordinate maintains high fat reserves under all circumstances, for the reasons already indicated. Thus the semigame model of dominant and subordinate behavior tends to support the hypothesized explanation for the observation of Ekman and Lilliendahl (1993).

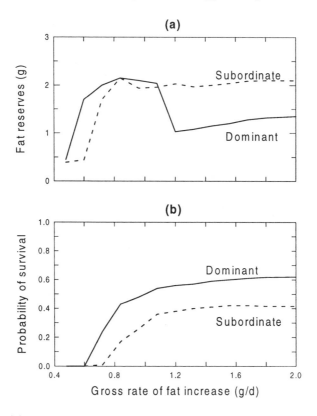

Figure 5.6 (a) Optimal fat reserves and (b) probability of survival over 120 d for dominant and subordinate members of a dyad of willow tits, as functions of daily rate of fat increase e_2 in the dangerous habitat H_2. From Clark and Ekman (1995).

5.3 Experiments suggested by the model

Our dynamic model of willow tit behavior suggests a number of experiments to test the predictions qualitatively. For example, we have already discussed the linear buildup of fat reserves during the day. The semigame model of section 5.2 suggests two removal experiments. First, if the dominant in a pair of birds is removed, the fat reserves of the subordinate should drop. Second, if the subordinate is removed, little change should be observed in the dominant. Field experiments conducted by Ekman and his colleagues confirmed both of these predictions (J. Ekman, personal communication).

Another prediction is that in areas where winter food supplies are scarce, both dominants and subordinates will maintain high fat reserves. Studies of willow tits in Finland support this prediction (K. Koivula, personal communication). Also, Gosler (1987) showed that subordinate great tits (*P. major*) in Britain hold larger reserves than dominants when food (beech mast) is scarce but not in mast years when food is abundant.

To summarize the results of sections 5.1–5.3, dynamic state variable models are useful in helping to understand the trade-offs involved in winter fattening strategies of song birds, taking into account a variety of ecological variables, including food supply, weather-induced metabolic costs, predation risk, and social interactions. These models generate predictions that agree qualitatively and quantitatively with field observations. Some of the predictions can be tested experimentally, and these tests have largely been successful. The models could be further refined by including other sources of environmental variability such as day-to-day, seasonal, and geographic variations in food availability.

5.4 Models of hoarding behavior

The model studied in section 5.1 indicates that a prolonged period of cold weather can have disastrous consequences for birds in winter, unless the birds can carry sufficient fat reserves to tide them over. But carrying fat reserves sufficient for long periods may be anatomically impossible. Even moderate fat reserves may have severe consequences in terms of predation risk. An alternative, which has been adopted by many bird species, is hoarding (caching, storing) of food supplies in winter (Andersson and Krebs 1978; Roberts 1979; Smith and Reichman 1984; Sherry 1985; Balda and Kamil 1989; Lucas and Walter 1991; Hitchcock and Houston 1994; Bednekoff and Balda 1996; Brodin and Clark 1997). Let us indicate how the basic model of section 5.1 can be modified to include hoarding. Later, we will discuss a more complex model.

We suppose that the willow tit starts the winter with a hoard z_0, and let $Z(t, d)$ denote the size of the hoard (measured in units of fat that can be obtained from it) in period t, day d. Thus $Z(t, d)$ becomes a second state variable in the model. We use H_3 to represent the activity of retrieving food from the hoard; predation risk is $\mu_3(x)$, and the retrieval rate is e_3. Unlike natural food sources, however, the hoard becomes depleted when used. The hoard is accessible only during daytime.

The fitness function $F(x, z, t, d)$ for the new model is defined in the usual way in terms of survival from t, d to $1, D + 1$. Then, in close analogy to eqs. 5.1–5.4,

$$F(x, z, T+1, D) = \begin{cases} 0 & \text{if } x < c_g \\ p_g & \text{if } c_g \leq x < c_b \\ 1 & \text{if } x \geq c_b \end{cases} \tag{5.8}$$

$$F(x, z, T+1, d) = \begin{cases} 0 & \text{if } x < c_g \\ p_g F(x - c_g, z, 1, d+1) & \text{if } c_g \leq x < c_b \\ p_g F(x - c_g, z, 1, d+1) & \\ \quad + p_b F(x - c_b, z, 1, d+1) & \text{if } x \geq c_b \end{cases} \tag{5.9}$$

$$F(x, z, t, d) = \max_i [1 - \delta\mu_i(x)]F(x_i', z_i', t+1, d) \text{ if } x \geq 0 \qquad (5.10)$$

where (subject to $0 \leq x_i' \leq x_{\max}$ and $z_i' \geq 0$)

$$x_i' = x + \delta e_i - \gamma\delta(m_0 + x) \qquad (5.11)$$

$$z_i' = \begin{cases} z & \text{if } i = 0, 1, 2 \\ z - \delta e_3 & \text{if } i = 3 \end{cases} \qquad (5.12)$$

Computational problems

Dynamic state variable models with two state variables are more complex than one-dimensional models for several reasons. (Things get rapidly worse with three or more state variables.) First is the problem of storing large arrays like $F(x, z, t, d)$ in the computer. Suppose we discretize x and z into N_x and N_z values. Then the array $F(x, z, t, d)$ will contain $N_x \cdot N_z \cdot T \cdot D$ entries. In the basic model of section 5.1 we used $N_x = 60$, $T = 50$ and $D = 120$, so that $F(x, t, d)$ had 360,000 entries. If we take $N_z = 100$, then $F(x, z, t, d)$ will have 36 million entries, which exceeds the memory of many desktop computers. Moreover, the program will need to compute each of these 36 million numbers, using eqs. 5.8–5.12. This ballooning of computational requirements in dynamic programming is called the "curse of dimensionality." Modern developments in computers have steadily reduced the impact of the curse of dimensionality, but they have not eliminated it. Typically, adding an additional state variable to a model increases computer requirements by a factor of 100 or more. Models that have more than three or four state variables rapidly become impossibly unwieldy.

To handle the hoarding model, we made several computational compromises. We reduced N_x to 20 and T to 10 periods per day (Clark and Ekman [1995] obtained useful results using only one period per day). Second, we stored $F(x, z, t, d)$ day-by-day, for $d = D, D-1, \ldots, 0$; this reduces storage requirements but does not change the number of calculations. Now the array for F has 200 N_z entries and requires 24,000 N_z calculations.

To obtain N_z, we took $z_{\max} = 6\,\text{g}$, enough food for five days of bad weather. To avoid unnecessary two-dimensional interpolation in both x and z, we used N_z discrete units of z and assumed that one time period of retrieving from the hoard would obtain one unit. The maximum daily retrieval rate was set at $e_3 = 2\,\text{gd}^{-1}$, so that the hoard can rapidly replenish depleted reserves. Therefore one unit equals $e_3/T = 0.2\,\text{g}$, and $N_z = 30$. Now the necessary computations are quite feasible.

These details have been discussed here to give you some insight into the decisions that have to be made in transferring the dynamic programming equations for a given empirical model into computer code. The more complex the model, the more critical these computational decisions become. Creativity and imagination are needed here as much as elsewhere in the modeling process.

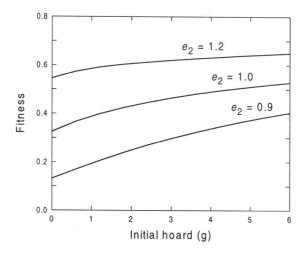

Figure 5.7 Fitness (overwinter survival probability) $F(1.0, z_0, 1, 1)$ as a function of initial hoard size z_0 for three levels of food availability e_2 in habitat H_2. A relatively small hoard of 6 g can have a strong influence on survival, particularly when e_2 is barely sufficient for average daily needs. The hoard permits birds to survive an occasional bout of cold weather.

How do we expect the new model to behave? If the hoard is critical for survival, we expect that the bird will exploit it only in desperate circumstances—namely, when its bodily fat reserves are too low to survive a bad night without using the hoard supply. But the main purpose of the new model is to check our intuition that relatively small hoards can have a large impact on winter survival rates under certain parameter combinations.

Figure 5.7 shows the effect of initial hoard size z_0 on fitness $F(x_0, z_0, 1, 1)$ (with $x_0 = 1.0$ g) for three values of e_2. For example, although $e_2 = 0.9$ exceeds average daily requirements, fitness is only 0.13 with no hoard. A hoard of 6 g increases fitness to 0.4, a threefold increase. (This assumes no predation risk while retrieving from the hoard, but this makes little difference.) The effect is somewhat less dramatic for larger values of e_2, and disappears entirely for $e_2 \geq 1.5$. Our intuition about the value of a relatively small hoard turns out to be correct (Hitchcock and Houston 1994). If e_2 is below the average daily requirement, a small hoard will have little or no effect on fitness. In this case, the bird's winter survival may depend on establishing a substantial store of seeds for winter use. Next we consider this possibility.

5.5 Long-term hoarding

Three parid species, willow tits, Siberian tits (*P. cinctus*), and boreal chickadees (*P. hudsonicus*) are winter residents in northern taiga forests. These species store large amounts during autumn, when food is abundant (Haftorn 1956; Brodin 1994). The stored food forms an important part of the winter diet (Jansson 1982). The exact locations of stored seeds do not appear to be remembered over winter (Hitchcock and Sherry 1990), but it has been suggested that stored seeds should increase the rate of food recovery in a bird's territory (Brodin and Clark 1997).

Brodin and Clark (1997) modeled the long-term hoarding behavior of willow tits; their model, which is considerably more complex than the model discussed in section 5.3, will be described briefly here. Previous models of hoarding behavior in parids (McNamara et al. 1990; Lucas and Walter 1991) dealt with short-term hoarding. For example, McNamara et al. assumed in most of their models that caches were retrieved on the same day they were stored, whereas Lucas and Walter assumed that most caches would be pilfered within five days.

Willow tits also use hoarded seeds in a short-term manner to be consumed within one or two days after caching (Brodin 1992, 1993). Short-term caches, the location of which seem to be remembered, allow the bird to carry less body fat. Long-term stores, on the other hand, are not memorized but may increase retrieval rates in the winter territory. These two types of hoarding occur in the same population of willow tits (Brodin 1993; Brodin and Ekman 1994). Dynamic state variable models are well suited to investigate how hoarding decisions depend on such factors as food availability, environmental conditions, and seasonal changes.

The long-term hoarding model employs three state variables:

$$X(d) = \text{body fat reserves (kJ)}$$
$$Y(d) = \text{remembered seeds (kJ), or short-term hoard}$$
$$Z(d) = \text{long-term stores (kJ)}$$

Here d denotes day in winter, $1 \leq d \leq D$. Because of the dimensionality problem, programming details were critical. The basic time period was taken as one 24-h day. Two seasons were considered, autumn ($d = 1, 2, \ldots, D_w - 1$) and winter ($d = D_w, \ldots, D$). Night-time metabolic costs $C(d)$ are higher in winter than in autumn. Similarly, the rate of energy gain while foraging is lower in winter than in autumn. Predation risk also increases from autumn to winter. Table 5.4 lists model parameters and their assumed values. During the day, the bird allocates its time among four activities:

A: forage and consume food that is found
B: forage and store seeds
C: perch in safe habitat
D: retrieve remembered seeds

(The symbols C and D in this list should not be confused with $C(d)$ and D_w introduced above. The meaning intended should be clear from the context.) Metabolic costs c_i and predation risks μ_i per day for each of these activities are listed in table 5.4. The decision variables are a_i = proportion of the day spent in each activity i = A,B,C,D with $a_A + a_B + a_C + a_D = 1$. The dynamics of state variables and decisions are shown in fig. 5.8.

Table 5.4. Parameters of the long-term hoarding model

Symbol	Meaning	Baseline value
D	Time horizon (days)	200
D_w	Start of winter (day)	100
$c_g(d)$	Metabolic cost on a normal night on day d (kJ)	$14.0 (d < D_w)$
		$29.0 \ (d \geq D_w)$
$c_b(d)$	Metabolic cost on an unusually cold night, day d (kJ)	$1.2 \times c_g(d)$
p_{ng}	Probability of a normal night	.90
p_{dg}	Probability of normal day-time weather	.95
γ_1	Proportion of stored seeds forgotten, per day	0.10
γ_2	Proportion of stored seeds lost, per day	0.013
γ_3	Proportion of long-term stores disappearing, per day	0.013
γ_4	Proporton of encountered food possible to store	1.0
$r_A(d, g)$	Energy gain (activity A), day d, good weather (g) (kJ/day)	$200 (d < D_w);$
		$45 (d \geq D_w)$
$r_A(d, b)$	Ditto, bad weather (b)	$0.7 \times r_A(d, g)$
r_B	Increase in stored supplies (Y) from hoarding (kJ/day)	$\gamma_4 r_A(d, w)$
r_D	Rate of recovering remembered seeds (kJ/day)	$r_A(d, g)$
$x_{\max}$	Maximum possible fat reserves (kJ)	65.0
$y_{\max}$	Maximum number of remembered seeds	2500
$z_{\max}$	Maximum number of forgotten seeds	25,000
ρ	Maximum possible effect on food supply by long-term stores	0.7
k	Coefficient making metabolic cost dependent on body mass	0.5
c_A	Metabolic cost, activity A (kJ/day)	12.0
c_B	Metabolic cost, activity B (kJ/day)	12.0
c_C	Metabolic cost, activity C (kJ/day)	10.0
c_D	Metabolic cost, activity D (kJ/day)	12.0
$\mu_A(d)$	Predation risk, activity A, day d (per day)	$0.0001 \ (d < D_w);$
		$0.0005 \ (d \geq D_w)$
$\mu_B(d)$	Predation risk, activity B, day d (per day)	$\mu_A(d)$
$\mu_C(d)$	Predation risk, activity C, day d (per day)	0
$\mu_D(d)$	Predation risk, activity D, day d (per day)	$\mu_A(d)/2$

a. From Brodin and Clark 1997.

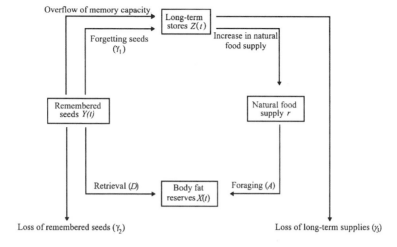

Figure 5.8 State variable dynamics for the long-term hoarding model. From Brodin and Clark (1997).

As shown in table 5.4, both days and nights may be "abnormal" (independently), and imply a reduction in food intake or an increase in metabolic cost, respectively. We let $W(d)$ denote the weather state, normal or abnormal, on day d; it is assumed that the bird knows the current state $W(d) = w$ and takes this into account in the day's allocation decision. The fitness function $F(x, y, z, w, d)$ is defined as the maximum probability of surviving from the beginning of day d to day $D + 1$, given $X(d) = x$, $Y(d) = y$, $Z(d) = z$, and $W(d) = w$. The function $F_n(x, y, z, d)$ is also defined in the same way but is evaluated at the start of night on day d. This is needed for the sequential coupling of day and night.

The terminal condition, analogous to eq. 5.8, is expressed as

$$F_n(x, y, z, D) = \begin{cases} 0 & \text{if } x < c_g(D) \\ p_{ng} & \text{if } c_g(D) \leq x < c_b(D) \\ 1 & \text{if } x \geq c_b(D) \end{cases} \tag{5.13}$$

Similarly, as in eq. 5.9,

$$F_n(x, y, z, d) = E_{c,w} F(x - c, y, z, w, d + 1) \tag{5.14}$$

Finally,

$$F(x, y, z, w, d) = \max_{\mathbf{a}}[1 - \mu_{\mathbf{a}}(d)] F_n(x'_{\mathbf{a},w}, y'_{\mathbf{a},w}, z'_{\mathbf{a},w}, d) \tag{5.15}$$

where $\mathbf{a} = (a_A, a_B, a_C, a_D)$ is the vector of decisions and $\mu_{\mathbf{a}}(d) = a_A \mu_A(d) + a_B \mu_B(d) + a_C \mu_C(d) + a_D \mu_D(d)$. The expressions for the values (e.g. $x'_{\mathbf{a},w}$) of the state variables at the end of day-time activities take account of the allocations **a**, the current weather w, and daytime metabolic costs c_i, as well as forgetting and pilfering of stores (Brodin and Clark 1997).

The role of long-term stores

As shown in fig. 5.8, long-term stores Z are built up from short-term stores that are forgotten or that exceed memory capacity. Long-term stores add to the natural food supply, and result in increased daily food intake while foraging (activity A), according to the formula

$$\text{Daily energy intake} = r_A(d, w) \cdot (1 + \rho \frac{z}{z_{max}})$$

where $r_A(d, w)$ denotes the daily energy gain from the natural food supply under weather conditions w; and z, z_{max} are the current and maximum long-term stores, respectively; and ρ is a coefficient that determines the effect of long-term stores on daily energy gain. During winter $(d \geq D_w)$ (table 5.4),

$$r_A(d, w) = \begin{cases} 45\,\text{kJ} & \text{(good day)} \\ 31.5\,\text{kJ} & \text{(bad day)} \end{cases}$$

Therefore,

$$\text{Daily energy intake} = \begin{cases} 45 - 75.5\,\text{kJ} & \text{(good day)} \\ 31.5 - 53.6\,\text{kJ} & \text{(bad day)} \end{cases}$$

for long-term stores ranging from zero to z_{max}. Daily (24-h) metabolic costs are

$$\text{Daily metabolic cost} = \begin{cases} 41\,\text{kJ} & \text{(good night)} \\ 47\,\text{kJ} & \text{(bad night)} \end{cases}$$

if the entire day is spent foraging. With zero long-term stores, bad nights result in depletion of fat reserves, particularly if days are also bad. Consequently, long-term stores are essential for survival. Figure 5.9 shows fitness—that is, survival—$F(x_0, y_0, z_0, w, d)$ in early and late winter $(d = 110$ or $190)$ as functions of short-term and long-term stores, y_0 and z_0, for $x_0 = 0.1x_{max}$. In early winter, fitness is nearly zero if z_0 is small, but increases rapidly, as z_0 is increased; for $z_0 = 0.5z_{max} = 12,500\,\text{kJ}$, fitness is nearly maximized. Total stores (short- plus long-term) are also important in late winter (fig. 5.9 c,d), although obviously less total stores are required.

These predictions concur qualitatively, but perhaps not quantitatively, with the results obtained from the simple hoarding model of section 5.3. In that model, a small hoard containing 2–3 d worth of food was sufficient to improve fitness considerably (see fig. 5.7). In the present model a much larger amount of long-term stores is required (12,500 kJ is enough for 265 d of food needs). The two models may seem to contradict one another, but the discrepancy results from the way in which stored food is assumed to be used.

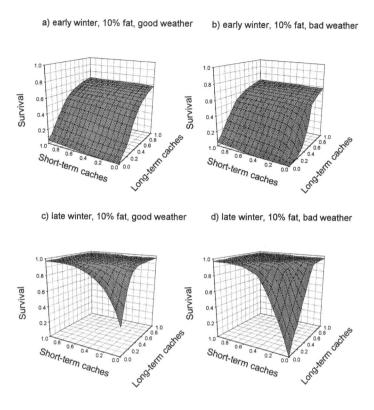

Figure 5.9 Fitness (winter survival) $F(x_0, y_0, z_0, w, d)$ for early winter ($d = 110$, top row) and late winter ($d = 190$, bottom row), under good (left column) and bad (right column) weather conditions, as functions of short (y) and long-term (z) caches. Caches are shown as portions of maximum levels y_{max}, z_{max}; also $x = X(d) = 0.1x_{max}$. From Brodin and Clark (1997).

In the earlier model of section 5.3, we assumed that the bird could locate and consume up to $2\,\mathrm{gd}^{-1}$ of hoarded seeds. In the current model, this rate of food intake in winter requires long-term stores $z \cong 23,000\,\mathrm{kJ}$. The earlier model tacitly assumes that the location of the hoard is known to the bird, whereas the current model assumes just the opposite. Field observations of willow tits suggest that the latter assumption is more realistic (Haftorn 1956; A. Brodin, personal communication).

Figure 5.10(a), obtained by forward iteration, shows typical trajectories of short- and long-term stores during winter. Short-term reserves $Y(d)$ rapidly accumulate to the memory capacity (2500 kJ). The bird continues to hoard seeds, which are forgotten and transferred to long-term stores $Z(d)$. Long-term stores accumulate to a peak of about 17,500 kJ by the beginning of winter and subsequently decline. As noted, these long-term stores are instrumental in aiding winter survival.

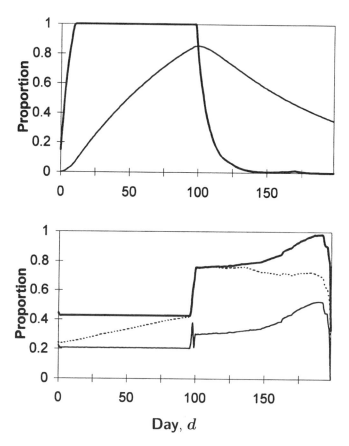

Figure 5.10 Mean state variables (1000 forward iterations): (a) short-term (thick line) and long-term (narrow line) stores as a proportion of maximum levels; (b) fat reserves at morning (narrow lines), evening (thick lines), and adjusted for changing day length at 60°N (dotted line). Onset of winter occurs at day 100. From Brodin and Clark (1997).

The model also predicts that the bird maintains much higher body fat reserves in winter than in autumn (fig. 5.10b). This agrees with the model of fattening strategy discussed in section 5.1. Another interesting prediction is that the level of body fat in winter should gradually increase, as long-term stores decline. The model, however, assumes that nights are equally long throughout the winter. In fact, of course, nights grow shorter as winter passes and reduce the needs for fat reserves. If the model is altered to include shortening nights, it is predicted that body fat levels will decrease during winter (Brodin and Clark 1997, fig. 3).

Discussion

Sections 5.1–5.5 illustrate the advantages of studying a series of models pertaining to the same general behavioral phenomenon. The simple models of

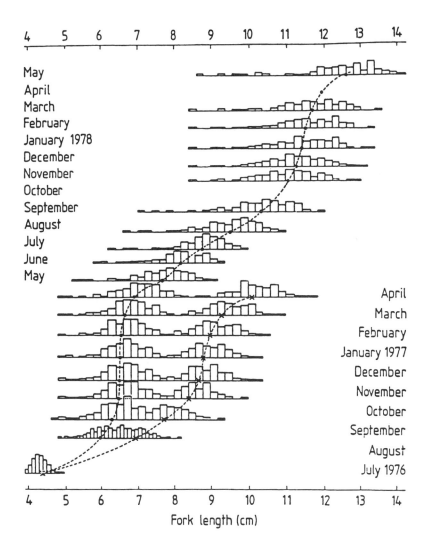

Figure 5.11 The development of bimodality in a cohort of sibling Atlantic salmon. From Thorpe et al. (1992).

fattening and hoarding strategy discussed in sections 1 and 3 help us to understand many aspects of the adaptations of willow tits to winter conditions. The more complex model of section 4, which required more than six months work, adds considerable biological realism and confirms the predictions of the simpler models.

Although the models discussed in this chapter pertain to the winter behavior of willow tits, they may provide insights into the forces affecting the behavior of other species faced with surviving difficult times. Some examples of such predictions are the preponderance of predation over starvation as the leading cause of death in winter, except in severe climates; the prediction that birds carry minimal fat reserve where food supplies are high, but need to carry larger reserves where food supplies are temporarily low or unpredictable; and the value of a relatively small hoard, provided that its location is remembered, versus the need for storing large amounts of food if location is not memorized. These predictions have implications for many other aspects of winter behavior and ecology such as territoriality, social interaction, migration strategy (whether or not to migrate), and local population dynamics in variable environments.

5.6 Overwinter foraging by juvenile salmon

Thorpe (1977) found that sibling Atlantic salmon (*Salmo salar* L.) have a very tight and approximately normal size distribution in May. Even when growing under uniform food and water conditions, by October the size distribution bifurcates into two nearly non-overlapping size distributions (fig. 5.11); this also happens in the field (e.g. Nicieza et al. 1991). The fish in the larger size class (the upper modal group, UMG) continue to feed over the winter and develop characteristics for life in seawater in the following spring (smolt metamorphosis) (Thorpe et al. 1980; Higgins and Talbot 1985; Metcalfe et al. 1986, 1988; Thorpe 1987). The fish in the smaller size class (lower modal group, LMG) become anorexic (i.e., lose appetite) during the winter and reduce foraging activity (Metcalfe and Thorpe 1992). This reduction in appetite commences in early autumn, independent of water temperature or food availability (Metcalfe et al. 1986), and lasts until spring, at which time water temperature rises and food becomes more plentiful. Food intake during the anorexic interval is insufficient to maintain weight or energy reserves (Gardiner and Geddes 1980; Higgins and Talbot 1985).

At the end of the winter, the anorexic fish resume active foraging and, in the lower latitudes, feed for an entire year (including their second winter of life) before undergoing smolt metamorphosis in the third spring of their lives. In upper latitudes, the seasonal pattern of foraging may be repeated for as many as seven or eight years. In general, fish that have relatively higher metabolic rates feed over the winter and smolt in the following spring, whereas fish that have lower metabolic rates lose appetite; thus, metabolic rate determines both the costs and growth potential of juvenile salmon (Metcalfe et al. 1995). The effects of the metabolic rate carry throughout the life of the fish, including the return size and timing of adults (Pentelow et al. 1933). The occurrence of bimodality has been demonstrated in some of the Pacific salmonids, particularly coho (*Oncorhynchus kisutch*), chinook (*O. tshawytscha*), and masu (*O. masou*) salmon and is likely in steelhead trout (*O. mykiss*).

Intuition about this behavioral pattern can be gained by recognizing that winter is a difficult time for fish. Food is scarce. Low water temperature slows the rate of metabolism (so that even if food is available, assimilated energy may be low) and reduces antipredator responses, especially from endothermic predators. During migration after smolt metamorphosis, fish are subject to size-dependent predation (bigger fish experience lower rates of predation). Fish that grow too slowly to successfully withstand size-dependent predation pressure during migration afterward follow developmental pathways that delay smolting. In response, predation risk over the winter is reduced by loss of appetite and minimal foraging.

Mangel (1994) developed a dynamic state variable model for separating LMG and UMG fish. In this chapter, we concentrate on questions concerning how LMG fish use fat reserves during the winter (Bull et al. 1996).

In previous sections, we showed that theory predicts that diurnal birds should generally carry large fat reserves in winter due to long nights (during which food is unavailable), harsh weather (creating an unpredictable feeding pattern) and the increased metabolic demands of homeothermy at cold temperatures. Furthermore, the response may differ for dominants and subordinates. This seasonal pattern has been found in many species. However, if animals reduce appetite because feeding is risky or difficult, as in the case of juvenile salmon, it is not clear how reserves should be used.

Previous work (Metcalfe and Thorpe 1992) showed that appetite at the commencement of overwinter anorexia in nonsmolting fish is sensitive to body lipids. Hence, we develop a dynamic state variable model using fat reserves as the state variable. We use the model to predict how responses to deviations from the programmed path of reserve loss should vary across the season. The model suggests experiments, which we summarize (Bull et al. 1996).

5.7 A state variable model for overwinter lipid dynamics

We envision a single state variable $X(t)$ that corresponds to the level of body lipids (measured in mg) at the start of day t within the anorexic interval ($t = 1$ is October 1; $t = 180$ is March 31, taken as the end of the winter, T). The juvenile fish that we model are nominally about 1 g at the start of the anorexic interval and have a maximum of 8% lipid, so that we constrain $X(t) \leq x_{max} = 80$ mg. If a fish forages for a fraction v of the 24-h period, the lipid dynamics are given by

$$X(t + 1) = X(t) - \alpha v - \alpha s(1 - v) + vq \qquad (5.16)$$

where α is active daily metabolic rate (mg/day), αs is resting metabolic rate (mg/day), so that s measures the reduction in metabolic needs of resting relative to activity, and q is lipids (mg) obtained during 24 h of foraging. We assume that fish die if their lipid level falls below a critical value x_c. Various experiments (C. Bull and N. Metcalfe, personal communication) suggest that this value is greater than zero, so we set it to $x_c = 10$ mg; however, because the state dynamics are linear, the precise choice of the critical value is not important.

The environment is characterized by the probability p_f that a fish can forage on a given day and a mortality rate m (day^{-1}), so that the probability that a fish survives a day in which it forages for a fraction v of the day is e^{-mv}. Increasing foraging effort increases the level of lipids, which increases survival by providing a buffer against the critical value of lipids. However, increasing foraging effort also increases the mortality risk. Thus, foraging involves a trade-off between starvation and predation risk. The trade-off should lead to foraging effort depending upon lipid level. For example, when lipid reserves are close to the critical amount, the animal must forage, regardless of the mortality risk, because otherwise death is almost certain. Similarly, toward the end of the season, if its lipid reserves are high enough, a fish may avoid foraging completely. Thus we anticipate that the optimal level of foraging effort depends upon both time and physiological state. To find it, we set

$$F(x,t) = \text{maximum Pr}\{ \text{ survival between } t \text{ and } T, \text{ given that } X(t) = x\}$$

(5.17)

which satisfies the end condition that $F(x,T) = 1$ if $x \geq x_c$. For $t < T$, $F(x,t)$ also satisfies the condition that $F(x,t) = 0$ if $x < x_c$ (the fish is dead from starvation) and, for $x \geq x_c$,

$$F(x,t) = (1 - p_f)F(x - \alpha s, t + 1)$$
$$+ p_f \max_v \{e^{-mv} F(x - \alpha v - \alpha s(1 - v) + vq, t + 1)\}$$

(5.18)

Here $\max_v$ indicates that the maximum is to be taken over the foraging effort $(0 \leq v \leq 1)$; the term in $\{\ \}$ takes into account predation risk and starvation risk. The solution of eq. 5.18 generates the survival probability and the optimal level $v^*(x,t)$ of foraging effort for every x and t.

Since the anorexic interval lasts 6 months, $T = 180$. Starved fish (Metcalfe and Thorpe 1992; Bull et al. 1996) lost about 1% of their wet weight in lipids over a three-week deprivation period. This corresponds to a change of 10 mg, so that we choose $21\alpha s = 10$. Thus, α is approximately $(0.5/s)$ mg/day. The additional parameters, q, s, m, and p_f are harder to estimate. Bull et al. (1996) assumed that the overwintering fish can meet its daily requirements by foraging for about 20% of a 24-h interval, so that $0.2\alpha + 0.8\alpha s = q$. They performed calculations for a wide variety of values for the other parameters, and the qualitative results are similar to those we report here. For presentation, we set $s = 0.2$, $p_f = .5$ and $m = 0.01$ day^{-1}.

The predicted optimal foraging effort increases as lipid levels decline and decreases as the end of winter approaches (fig. 5.12). From this, we predict that fish in good condition will forage little or not at all, so that there will be an initial decline in weight early in the winter. We also predict that fish in poor condition will forage with great intensity regardless of the time of year. The results in fig. 5.12 can be summarized as a boundary curve that separates regions in which we predict the fish will forage from those in which we predict

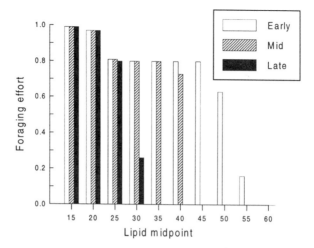

Figure 5.12 Predicted foraging effort as a function of state (lipid midpoint) early ($t = 6$: white), mid ($t = 66$: striped) or late ($t = 126$: black) season. Note that for lipid levels near the critical value, the prediction is that fish will forage, regardless of time of year. However, as lipid levels increase, foraging effort decreases and ultimately stops in late season or midseason.

it will not forage (fig. 5.13). The boundary curve approach, however, does not provide any information about how much foraging effort should be expended, if it is predicted that the animal will forage.

The boundary curve is a dynamic set point for the state value at which the fish will forage. Jobling and Miglavs (1993) report a similar role of fat in regulating feeding behavior in Arctic char, although they envision a fixed set point, rather than a dynamic trajectory. One sensitivity analysis is to ask how the qualitative features of a boundary curve, as in fig. 5.13, changes with changing assumptions. For example, (1) if the terminal condition is changed so that $F(x, T)$ is an increasing function of x, the declining curve is not observed (since it is beneficial to acquire additional lipids toward the end of the winter); (2) when the mortality rate $m = 0$, the threshold state becomes independent of time; however, if m is as small as 0.001, the pattern is maintained; (3) if $p_f = 1$, the pattern disappears, and the fish forages in mid- or late season only if the state is near the critical value; however, the pattern is maintained for $p_f = .7, .4$ or $.3$; or (4) the pattern is maintained—although the particular quantitative relationships change—if $s = 1.0$, $s = 0.2$, or if foraging for 40% or 60% of the day is required to meet metabolic needs.

We can use forward iteration to predict how fish will respond to deprivation at different times in the season. Metcalfe and Thorpe (1992) and Bull et al. (1996) used deprivation periods lasting three weeks, and Bull et al. started the deprivation periods early in the winter ($t = 6$), midwinter ($t = 66$), or late in the winter ($t = 126$). We used forward iteration and computed the average foraging effort for 14 days after the deprivation period (table 5.5),

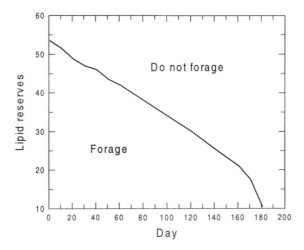

Figure 5.13 The results of the model can be summarized as a boundary curve that separates the time-state variable plane into regions in which the fish will forage (although the amount of foraging is not specified) and regions in which it will not forage. Sensitivity analysis shows that decreasing the value of m, decreasing the value of p_f, or increasing the fraction of the day needed to meet requirements will move the boundary upward. From Bull et al. (1996).

Table 5.5 Predicted foraging effort of control and deprived fish after three deprivation periods

Time of deprivation	Average foraging effort for 14 days after the end of the deprivation period	
	Control	Deprived
Early	0.13	0.33
Mid	0.11	0.13
Late	0.06	0.06

using the empirical value of lipid content at the start of the deprivation period ($x = 51$, 45, or 35, respectively). The two main predictions from table 5.5 are (1) that control fish will decrease foraging effort during the season and (2) that experimental fish (deprived of food for three weeks) will, following the deprivation period, have a highly enhanced foraging effort early in the season, a slightly enhanced foraging effort in midseason, and no change in foraging effort late in the season.

5.8 Empirical test of the predictions

Bull et al. (1996) tested these ideas. The fish were individually marked and divided into two groups of 20. Control fish experienced ambient water temper-

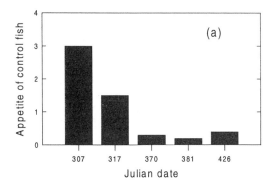

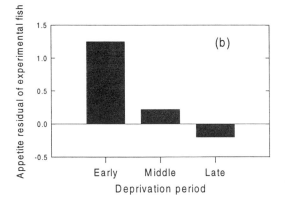

Figure 5.14 (a) As predicted by the theory (vertical comparison, table 5.5), the appetite of control fish decreased during the winter. (b) The appetite of experimental fish, relative to the control fish, following a deprivation period, depended upon the time at which deprivation occurred. Appetite was significantly higher for experimental fish than control fish following deprivation in early season ($p < .01$) and was greater following early deprivation than either mid or late-season deprivation ($p < .001$); further details and analysis are in Bull et al. (1996).

atures, a simulated natural photoperiod and excess food (dispensed every 10 min by an automatic feeder). They were kept in these conditions throughout the experiment, except when subjected to feeding trials.

Experimental fish were subjected to three separate three-week periods of food deprivation commencing on October 3 (early), December 5 (mid) and February 6 (late), respectively. Lipid level was estimated from a combination of morphometric measurements (Simpson et al. 1992; Bull et al. 1996). At the end of each deprivation period, the appetite of all fish was scored in response to five live bloodworms. The scores for different responses were: no response 0 ; orientation 1; turn back after initially moving toward the prey 2; miss bloodworm 3 ; ingest but subsequently reject bloodworm 4; consume bloodworm 5 (Metcalfe et al. 1986). Each fish received a minimum of 20 frozen bloodworms following each trial and on days between trials to ensure that they received ad lib rations. Individual daily appetite scores for experimental fish during refeeding were calculated as residual values from the mean value for control fish on that day; elevated or reduced appetites were indicated as positive or negative values, respectively.

We make two comparisons between the theory and experiments. First, from the vertical comparisons in table 5.5, we predict that the appetite of control fish will decline during the season (fig. 5.14a). Second, following deprivation early in the season, we predict that experimental fish will exhibit a marked elevation in appetite when compared to controls during the first 14 days of refeeding. However no such effect should be apparent after periods of deprivation in mid- or late season (fig. 5.14b). Note that if the fish were using a fixed, rather than dynamic, threshold for foraging, we would predict identical responses following deprivation, regardless of the time of the season. Further details and analysis of the results are found in Bull et al. (1996).

The theory predicts and experiments confirm that the fish are following seasonal trajectories toward a low target level of lipids early in the spring. Lipids can be restored as water temperatures increase (Cunjak and Power 1987; Simpson 1993). The different responses of appetite to deviations from this lipid trajectory (fig. 5.14b) suggest that the fish facultatively respond to their individual projection of whether they will be higher or lower than the target level at the end of the anorexic interval. The amount by which individuals deviate from the seasonal trajectory during the midwinter period will depend upon individual variation in metabolic rate and food acquisition ability. Thus, we predict that fish that have higher metabolic rates or poor foraging capabilities will require higher levels of insurance to guard against starvation late in the winter. In this manner, the anorexic fish maximize their probability of survival over the winter.

6

Avian Migration

Techniques and concepts used in chapter 6

Rate-maximizing models (6.1, 6.2)
Stochastic environments (6.3, 6.4)
Parameter tuning (6.4)
Monte Carlo forward iteration (6.4)
Sensitivity testing (6.4)
Life-history consistency (6.4)

The migration of birds is a complex form of behavior, affected by many environmental factors, such as spatial and temporal variations in food availability, predators, temperature, winds, and breeding opportunities (Ens et al. 1994). Until recently, most optimization models of migration have dealt with isolated aspects like optimal migration speed (Alerstam 1979; Liechti 1995; Hedenström and Alerstam 1995) or optimal fuel loads (Weber et al. 1994). The dynamic state variable approach allows one to include many factors in a single model (Weber and Houston 1997; Weber et al. 1998; Farmer and Wiens 1998, 1999; Clark and Butler 1999). For a species whose life cycle has been thoroughly studied, such a model can help to organize and evaluate our understanding of the way various selective forces are involved in determining migratory behavior. Then the relative importance of different environmental factors can be assessed by computer experiments and can lead to insights that would be difficult or impossible to obtain from field experiments.

In this chapter we describe a dynamic state variable model for the spring migration of western sandpipers (*Calidris mauri*), an abundant New World shorebird that migrates between tropical and subtropical wintering areas to its breeding grounds in northwest Alaska and Siberia, and refuels at various sites along the Pacific flyway. Since this model is rather complex, we will

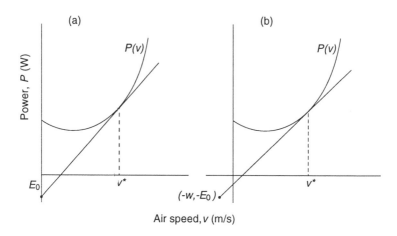

Air speed, v (m/s)

Figure 6.1 The power curve $P(v)$ for flight. (a) "Optimal" flight speed v^* in still air, defined as the speed that minimizes the total time of migration, is determined by eq. 6.2. This is obtained graphically as the point of tangency of the straight line drawn from point $(0, -E_0)$ to the power curve $P = P(v)$, as shown. (b) Graphical determination of the optimal flight speed v^* in the presence of a tailwind w. (For the case of a headwind $w < 0$, the point from which the tangent is drawn lies to the right of the y-axis.) These diagrams are analogous to those of Charnov (1976).

build up to it slowly by considering some simple models pertaining to flight energetics and related topics. For a survey of research and models of optimal migration see Alerstam and Hedenström (1998).

6.1 The energetics of bird migration

A power curve for the flight of a particular bird species (fig. 6.1) relates power input $P = P(v)$ to air speed v. Such power curves can be derived from the theory of flight mechanics and have also been measured in wind-tunnel experiments for some species (Pennycuick 1989; Norberg 1990). These curves are generally U-shaped, as shown.

Imagine a migrating bird, interspersing flights at speed v (m/s) with refueling stopovers that provide a constant rate of refueling E_0 (J/s) (Alerstam 1991). To calculate the average speed of migration for a series of such flights and stopovers, we let T denote the total flying time. Then, the distance traveled is $D = Tv$ and the total energy used in flying is $E = TP(v)$. The time spent refueling is $T_{\text{refuel}} = E/E_0 = TP(v)/E_0$. Hence the average migration speed is

$$v_{\text{mig}} = \frac{D}{T + T_{\text{refuel}}} = \frac{Tv}{T + TP(v)/E_0} = \frac{vE_0}{E_0 + P(v)} \qquad (6.1)$$

The flight speed $v = v^*$ that maximizes average migration speed (equivalent to minimizing the time required for migration) is obtained by setting $dv_{\text{mig}}/dv = 0$ and solving for $v = v^*$. This gives[1]

$$P'(v^*) = \frac{E_0 + P(v^*)}{v^*} \tag{6.2}$$

The solution v^* to eq. 6.2 can be obtained graphically, as shown in fig 6.1(a). Two qualitative predictions follow from this model:

Prediction #1: The flight speed of migrating birds will be higher than the speed v_0 that minimizes power use, but lower than the maximum attainable speed.

Prediction #2: The flight speed of migrating birds is an increasing function of the refueling rate.

This calculation presumes time minimization for a migrant that makes repeated refueling stopovers. Some migrants such as knots (*Calidris canutus*) migrate in one or two long flights between fixed staging areas (Piersma 1994). In such cases optimal flight speed might be defined in terms of minimizing total energy used for migration: minimize $E = TP(v)$, given $Tv = D$, where D is the given migration distance. Substituting $T = D/v$ in the expression $E = TP(v)$, we see that this is the same as: minimize $DP(v)/v$. Using the same rule $f'/g' = f/g$ as before, we see that (you should check this) the energy-minimizing flight speed v_E^* is the solution of the equation

$$P'(v) = \frac{P(v)}{v} \tag{6.3}$$

This is the special case of eq. 6.2 obtained by setting $E_0 = 0$. From fig 6.1(a), we see that $v_E^* < v^*$. In other words, the energy minimization criterion implies a lower flight speed than the time minimization criterion. Bruderer and Weitnauer (1972) (cited in Ens et al. 1994) showed that common swifts (*Apus apus*) maintained a speed during sleeping flights approximately equal to the power minimizing speed v_0 and used flight speeds close to the energy minimizing speed v_E^* while migrating.

---

[1] Note that, by calculus, maximization of a ratio $f(x)/g(x)$ implies that $(f/g)' = (f'g - fg')/g^2 = 0$, which implies that $f'g - fg' = 0$. The latter condition can be rewritten as $f'/g' = f/g$. Equation 6.2 is obtained from eq. 6.1 using this rule:

$$\frac{E_0}{P'(v)} = \frac{vE_0}{E_0 + P(v)}$$

which implies eq. 6.2.

Effect of winds

This calculation tacitly equates ground speed with flight speed v, which of course is only valid in the absence of winds. To include winds, first consider a tailwind of speed w, so that ground speed is $v + w$. Thus $D = T(v + w)$, and now the average migration speed is

$$v_{\text{mig}} = \frac{(v + w)E_0}{E_0 + P(v)} \tag{6.4}$$

Maximization of this expression implies that

$$P'(v) = \frac{E_0 + P(v)}{v + w} \tag{6.5}$$

Figure 6.1(b) illustrates the graphical solution of this equation for the case $w > 0$ (i.e., a tailwind). To explain this graph, first note that the expression on the right side of eq. 6.5 is equal to the slope of the line joining the points $(-w, -E_0)$ and $(v, P(v))$. Equation 6.5 says that this must be the same as $P'(v) =$ slope of the tangent line to the curve. In other words, to satisfy eq. 6.5, $v = v^*$ must be chosen so that the line joining the two points is tangent to the curve, exactly as shown in fig 6.1(b). This construction, analogous to the marginal value theorem (Charnov 1976; Stephens and Krebs 1986), allows us to obtain the solution v^* to eq. 6.5 graphically. Moreover, the graphical solution leads immediately to qualitative predictions of the way the parameters w and E_0 affect the optimal flight speed v^*. For example, stronger tailwinds w move the point $(-w, E_0)$ to the left, and this causes v^* also to move to the left (check this in fig 6.1(b)).

Prediction #3: Migrating birds will decrease their flight speed in tailwinds, relative to windless conditions. The stronger the tailwind w the slower the flight speed v^* (but ground speed $w + v^*$ may increase as w increases).

What about headwinds $w < 0$? We leave it to you to formulate and argue the prediction for this case. By the way, any bird obviously has a finite maximum possible flight speed v_{max}; this could be shown in fig 6.1 by drawing the power curve $P(v)$ with a vertical asymptote, as v approaches v_{max}. What should the bird do when headwinds $-w$ are greater than v_{max}?

Next, consider the case of a crosswind—see fig. 6.2, in which $\overrightarrow{w}$ is the wind vector and $\overrightarrow{v}$ the airspeed vector. We assume that the bird compensates for wind, if possible, by adjusting $\overrightarrow{v}$, so that the resulting ground speed vector $\overrightarrow{v}_g = \overrightarrow{v} + \overrightarrow{w}$ points along the desired ground track to the migration target, as shown. This implies that the bird flies to the target in a straight line. We ignore the earth's curvature, which is not a minor consideration for long range

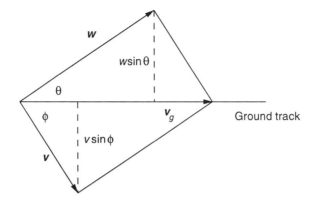

Figure 6.2 Wind compensation: it is assumed that the groundspeed vector $\boldsymbol{V_g} = \boldsymbol{v} + \boldsymbol{w}$ points along the desired migration track. Here, $\boldsymbol{w}$ is the wind vector, and $\boldsymbol{v}$ the bird's airspeed vector. (Boldface symbols represent vectors here.)

migrations. See Alerstam (1990, section 4.8) for further discussion. If θ and ϕ denote the angles between $\overrightarrow{w}$ and $\overrightarrow{v}$ and the track, then

$$v \sin \phi = w \sin \theta \qquad (6.6)$$

where w is wind speed and v is air speed of the bird. Equation 6.6 is the wind-compensation condition, that specifies the heading ϕ that the bird uses to compensate for wind drift. Since $|\sin \phi|$ must be < 1, compensation is possible only if

$$v > w \sin \theta$$

Then ground speed v_g is given by

$$
\begin{aligned}
v_g &= v \cos \phi + w \cos \theta \\
&= \sqrt{v^2 - w^2 \sin^2 \theta} + w \cos \theta
\end{aligned}
\qquad (6.7)
$$

(using eq. 6.6 and the identity $\sin^2 \theta + \cos^2 \theta = 1$).

For the time minimization model, $D = T v_g$, and the average migration speed is (see eq. 6.4)

$$
v_{\text{mig}} = \frac{v_g E_0}{E_0 + P(v)} = \frac{\sqrt{v^2 - w^2 \sin^2 \theta} + w \cos \theta}{E_0 + P(v)} \cdot E_0
\qquad (6.8)
$$

Maximization of this expression implies that

$$
P'(v) = [E_0 + P(v)] \frac{v}{v_g \sqrt{v^2 - w^2 \sin^2 \theta}}
\qquad (6.9)
$$

where v_g is given by eq. 6.7. You can check that eq. 6.9 reduces to eq. 6.5 for a tailwind ($\theta = 0$) or a headwind ($\theta = 180°$). There is no tangent-line construction similar to fig 6.1(b) for the solution v_θ^* to eq. 6.9, but it can be shown that

Prediction #4: $v_0^* < v_\theta^* < v_{180°}^*$ where v_0^* and $v_{180°}^*$ are the optimal air speeds for tailwind and headwind, respectively. v_θ^* also increases monotonically as the wind's angle θ increases from 0 to 180°.

The various predictions discussed in this section are summarized here:

1. Migrating birds use flight speeds that are greater than the power-minimizing speed v_0.
2. Increased refueling rates imply increased flight speeds for migrating birds.
3. The flight speeds of migrating birds are greater in headwinds, and smaller in tailwinds, than in zero wind conditions.
4. Migrating birds compensate for wind drift by adjusting their flight speed vector.

A number of important assumptions underlie this theory of optimal flight speeds for migrating birds. The models assume, for example, that natural selection will tend to minimize either the time spent or energy utilized on migration. Suggested justifications for these assumptions are that either migration or obtaining food is dangerous. More general theories would take into account the multiple trade-offs involved in migration decisions; we discuss such theories later in this chapter.

The wind-compensation model assumes that winds remain constant and that birds can fully compensate for wind direction. Both assumptions are discussed by Alerstam (1979, 1990) and Liechti (1995). When winds are variable, the optimal track may deviate considerably from a straight line (or great circle). For example, many birds that migrate in the fall from eastern North America to South America head out into the open Atlantic and thereby prepare to take advantage of different prevailing winds at different latitudes en route (Alerstam 1990). Alerstam also presents evidence showing that migrating birds compensate well for side winds when flying over land but tend to drift with the winds, to some extent, when over the sea.

6.2 Optimal fuel loads

Now we ask, how frequently should a migrating bird stop to refuel, assuming that suitable foraging habitat is continuously available en route? By stopping frequently, the bird could fly with minimal fat reserves and thereby reduce flight cost. Frequent stops, however, involve costs associated with landing and takeoff, finding local food patches, and so on. There is a similar and simultaneous problem for carrying water, but we ignore it here for simplicity.

Flight-range models

Several different models of maximum flight range as a function of fuel (fat) load have been proposed (Pennycuick 1975; Rayner 1988; Alerstam and Lindström 1990; Weber and Houston 1997). They are all based on an equation of the form

$$dx = J(x)\,ds \qquad (6.10)$$

where x denotes fuel load (measured in kJ of energy content, for example) and dx denotes the amount of fuel used to fly distance ds in still air. Here $J(x)$ is a given function, equal to fuel consumption per unit distance. Fuel consumption increases with increasing fuel load, that is, $J(x)$ is an increasing function of x. Different models are based on different functional forms for $J(x)$. In Pennycuick's model, derived from the physics of flight, the fuel consumption function has the form

$$J(x) = a(1 + bx)^{3/2} \qquad (6.11)$$

where a and b are species-specific constants.

In a flight beginning with fuel x_0 and ending with remaining fuel x_1, the total air distance D flown, is given by

$$D = \int_0^D ds = \int_{x_1}^{x_0} \frac{dx}{J(x)} \qquad (6.12)$$

In particular, the maximum *flight range*, given x_0, is given by

$$Y(x_0) = \int_0^{x_0} \frac{dx}{J(x)} \qquad (6.13)$$

Thus $dY/dx_0 = 1/J(x_0)$, so that flight range Y increases at a decreasing rate, as the initial fuel load x is increased (see fig. 6.3).

Now we consider the optimal fuel load for a bird migrating over continuous habitat suitable for feeding; song birds that migrate over North American forests would be one possible example. If x denotes the fuel load obtained at each stopover, then the bird flies distance $D = Y(x)$ before stopping to refuel. We ignore winds, for simplicity.

Assume that the time required to refuel to level x is $T_{\text{refuel}} = c + x/E_0$, where c and E_0 are positive constants, E_0 represents the refueling rate, and c is a fixed time-cost associated with the time used to land and take off again, locate a food source, etc. We also assume a constant flight speed v, which implies that the time needed to complete one flight is $T_{\text{flight}} = D/v = Y(x)/v$. Hence, the average migration speed is

$$v_{\text{mig}} = \frac{D}{T_{\text{flight}} + T_{\text{refuel}}} = \frac{Y(x)}{Y(x)/v + c + x/E_0} \qquad (6.14)$$

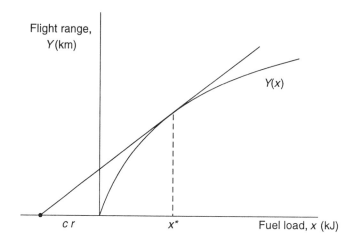

Figure 6.3 Graphical determination of the optimal departure fuel load x^*, as given by eq. 6.15 (in anticipation of later models we used r in place of E_0 here).

By differentiation, we obtain the equation

$$Y'(x) = \frac{Y(x)}{x + cE_0} \tag{6.15}$$

for the optimal take-off fuel load $x = x^*$. The graphical solution of eq. 6.15 is shown in fig. 6.3. This leads to the following predictions:

Prediction #5: The optimal departure fuel load is an increasing function of the refueling rate E_0, and of the fixed stopover time c.

Prediction #6: The optimal departure fuel load is independent of flight speed.

Prediction 5 says, for example, that if the fixed stopover cost c is small, then, the bird should stop frequently and fly with minimal fuel load. (The fact that you probably do not operate your car in this way suggests that you consider the time spent stopping at the gas station more important than the slightly increased mileage you would get by maintaining the gas tank nearly empty.) Prediction 6 implies that optimal departure loads do not vary with prevailing winds, although of course the distance between stopovers would depend on the winds. Empirical testing of these predictions seems lacking, perhaps because of the difficulty in finding situations in which the assumptions of the model are valid.

For example, the assumption of continuous habitat, allowing birds to stop and refuel at any point, would be unrealistic in many cases such as migration over water or desert areas. If refueling sites are limited, optimal fuel loads would account for the need to complete the flight to the next site. This question is discussed in section 6.3.

Weber and Houston (1997) and Weber et al. (1994) suggest that the form of the assumed fuel-consumption function $J(x)$ may strongly affect quantitative predictions of optimal departure loads. Clark and Butler (1999) found that $J(x)$ can also affect predicted migration schedules of western sandpipers; see section 6.4.

From the point of view of this book, however, the principal limitation of the type of models discussed in sections 6.1 and 6.2 is the restricted view of the operation of natural selection built into the assumed optimization currencies of time or energy minimization. Therefore in the next section, we reconsider the departure load problem from a much wider viewpoint, using the dynamic state variable approach. In the last section of the chapter, we discuss an empirical model of shorebird migration that incorporates ideas from sections 6.1–6.3.

6.3 A dynamic state variable model of single-flight migration

In this section, we develop a migration model based on the following assumptions:

1. At the initial time, a bird is located on its wintering ground.
2. Predation risk on the wintering ground is mass-dependent and also depends on foraging intensity.
3. The bird migrates to its breeding ground in a single flight.
4. Winds encountered on migration are stochastic.
5. Reproductive success depends on the date of arrival on the breeding ground and on the energy reserves that the bird has on arrival.

(At this stage you might find it a worthwhile project to develop your own dynamic state variable model based on these assumptions. Then you could compare your model and ours. As a less ambitious exercise, we suggest that you try to identify the decisions and trade-offs involved in this situation.)

Among complications that we do not include in our model are the molting decision, return migration strategy, and social interactions such as flocking and competition for breeding territories or mates. We also assume a fixed air speed v, and ignore the question of choosing an optimal flight speed.

We use one day as our basic time interval and choose a time horizon T that encompasses the remainder of the winter, the migration, and the breeding season. We do not model breeding decisions specifically but assume a terminal fitness function $\Phi(x, t)$ for a bird that arrives on the breeding ground on day t with energy reserves x. Thus, energy reserves $X(t)$ are one state variable in the model. A second, environmental state variable $Z(t)$ denotes a weather/winds "cue" on the wintering ground. In the simplest case,

$Z(t) = z$ would specify exactly what wind vector $\overrightarrow{w}_z$ the bird would encounter on migration, if it were to depart immediately on day t. More generally, $Z(t) = z$ determines the conditional probability distribution $p(\overrightarrow{w}|z)$ for winds encountered in migration. We do not consider other aspects of weather, such as cloud cover or precipitation.

The decisions on day t are (a) departure (a no-yes, irreversible decision) and if the decision is not to depart, (b) foraging effort e ($0 \leq e \leq 1$); $e = 1$ means that the entire day is spent foraging. Thus, the state dynamics for $X(t)$ on the wintering grounds are given by

$$X(t+1) = X(t) + er - c \qquad (6.16)$$

where r denotes the daily rate of energy gain while foraging, and c denotes the daily metabolic cost. Also $X(t)$ is constrained by $0 \leq X(t) \leq x_{\max}$. Daily predation risk, which depends on the day's effort and the bird's mass, is denoted by $m(x, e)$.

The dynamics of $X(t)$ for the migration flight are influenced by the wind vector $\overrightarrow{w}(t)$, which we assume is constant for the whole flight. Given $\overrightarrow{w} = \overrightarrow{w}(t)$, the ground speed v_g is given by eq. 6.7. The duration of the flight is $\Delta t = D/v_g$ (where D is the ground distance), and the total air distance is $D_{\mathrm{air}} = v\Delta t = Dv/v_g$.

Given the departure load $x = X(t)$ on day t, the bird's flight range is $Y(x)$, as in eq. 6.13. The bird fails to complete the flight if $Y(x) < D_{\mathrm{air}}$, in which case it is assumed that it perishes in the attempt. Otherwise, it arrives on the breeding ground at time $t + \Delta t$, and has remaining energy reserves x' determined by (see eqs. 6.12, 6.13)

$$D_{\mathrm{air}} = \int_{x'}^{x} \frac{dx}{J(x)} = Y(x) - Y(x') \qquad (6.17)$$

Expected reproductive success upon reaching the breeding ground on day $t' + \Delta t$ with reserves x' is $\Phi(x', t + \Delta t)$.

For the dynamic programming equation, let $F(x, z, t)$ denote the fitness function of a bird on the wintering ground, and let $V_1(x, z, t)$ and $V_2(x, z, t)$ denote the fitness of a bird that decides to depart or remain on the wintering ground, respectively.

$$V_1(x, z, t) = E\{\Phi(x', t + \Delta t)|z\}$$

$$= \sum_j p(\overrightarrow{w}_j|z)\Phi(x'_j, t + \Delta t_j) \qquad (6.18)$$

$$V_2(x, z, t) = \max_{0 \leq e \leq 1}[1 - m(x, e)]E\{F(x + er - c, z', t + 1)\} \qquad (6.19)$$

where we have assumed a discrete distribution $\{\vec{w}_j\}$ for the wind vector $\vec{w}$ and where x'_j and Δt_j are calculated, as explained above. Finally

$$F(x, z, t) = \max\{V_1(x, z, t), V_2(x, z, t)\} \tag{6.20}$$

Our purpose in developing this single-flight migration model was simply to fix ideas for the dynamic modeling of bird migration. We did not attempt to code the model or to obtain predictions (see Weber et al. 1998 for a related model).

6.4 Spring migration of western sandpipers

The western sandpiper (*Calidris mauri*) is a small New World wader that has a population of several million birds (Wilson 1994). It breeds on the coastal tundra plains of western Alaska and eastern Siberia and winters along beaches and mudflats from California to Peru and from the Carolinas to Venezuela. In April and early May, most birds migrate along the Pacific flyway, and employ discrete stopover sites located on the deltas of major rivers, where food is plentiful. The breeding season is short, starting in mid-May. Southward migration begins in late June and proceeds in two waves, adults followed by juveniles.

Clark and Butler (1999) developed a dynamic state variable model for the spring migration of western sandpipers that had two objectives. First, they wished to know whether existing information about the ecology of the species, combined with a fitness maximization hypothesis, was sufficient to explain the observed migration pattern. If so, it was hoped to use the model to assess the fitness consequences of potential changes in the species' environment such as loss of stopover habitat or climatic changes that affect wind patterns along the migration route. In this section, we describe the model and discuss its predictions. The question of environmental changes will be taken up in chapter 8.

The main question that arose while developing the model concerned the trade-offs that could influence the timing of migration for this species. First, winds along the Pacific flyway are strong and variable in spring; therefore Clark and Butler (1999) decided to make variable winds a major feature of the model. This in turn required considering the birds' ability to detect and utilize favorable winds. Because winds are unpredictable, there is also a benefit to being located as far north as possible early in the spring, so as to be in position for the next flight. Unless there is some benefit to staying at southern latitudes, the model should predict that birds migrate north as soon as possible, provided that food is available there. Western sandpipers apparently do not do this; for example, they are absent from the Fraser River delta (latitude 49° 5′ N) in winter, even though the related Dunlin (*C. alpina*) spends the winter there in some numbers.

These considerations led to formulating the hypothesis that in winter western sandpipers face lower predation risks at tropical or subtropical sites than they would at more northern sites. Several facts lend support to this hypothesis. First, avian predators such as peregrine falcons (*Falco peregrinus*) are common at the North American sites in winter; whether such predators are equally common at tropical sites is unknown. Second, harsh winter conditions can result in sporadic reductions in food availability at high latitude sites such as the Fraser River delta, which occasionally freezes over for a period of one to several weeks in winter. Shorebirds that winter at such a site need to carry sufficient fat reserves either to survive periods of low food supply or to move to a more southerly site.

Carrying large fat reserves, however, reduces birds' maneuverability (Metcalfe and Ure 1995) and probably increases their vulnerability to predators. It is known that wintering western sandpipers and other small shorebirds maintain low levels of fat reserves in both the New and Old Worlds (Sutherland 1991; Zwarts et al. 1990). This has usually been attributed to low resource levels at these sites, but this explanation leaves open the question why the birds fly so far just to encounter food shortages! The possibility that maintaining low fat reserves is a deliberate predator-avoidance strategy has been raised by Lima (1986).

It is possible that limitations or abundance of food supply at certain locations and times of the year also affect the migration strategies of western sandpipers. Lacking any data on this topic, Clark and Butler decided not to consider variations in food supply in the model.

The model covers the spring migration of females from Julian dates 90 (March 31) through 151 (May 31). As before, time units are one day. The stopover sites (s) considered are 0: Northwest Mexico (ME), 1: San Francisco Bay (SF), 2: Fraser River delta (FR), 3: Stikine River delta (ST), and 4: Copper River delta (CR); see fig. 6.4. It is assumed that birds proceed through these sites in order, although if conditions are favorable a bird may continue without spending any time at a given site. Birds are also allowed the option of terminating their migration at any time and of returning to the winter site. More complicated movements back and forth between different stopover sites were not considered; similarly, the possibility of stopping briefly at other locations, for example in inclement weather, was not included in the model.

State variables are $X(t)$ = energy reserves (kJ) and $S(t)$ = current site, with $0 \leq X(t) \leq x_{\max}$ and $S(t) = 0,1,2,3$, or 4. The lean body mass of western sandpipers averages about 22 g and the heaviest birds observed at stopovers weigh about 38 g. The difference is accounted for by fat and protein reserves. The energy content of fat is about 39 kJ/gm, and protein is considerably less. This leads to $x_{\max} \cong 600\,\text{kJ}$. The feeding rate at each stopover site was treated as a constant, $\Delta e_1 = 39$ kJ/d, except that the two northern sites CR and ST were assumed unavailable before date 110 because of possible ice conditions. This feeding rate is the maximum possible rate of fat gain estimated for small shorebirds (Lindström 1991; Butler et al. 1997); the model does not specifically consider feeding strategies at stopover sites (see Weber

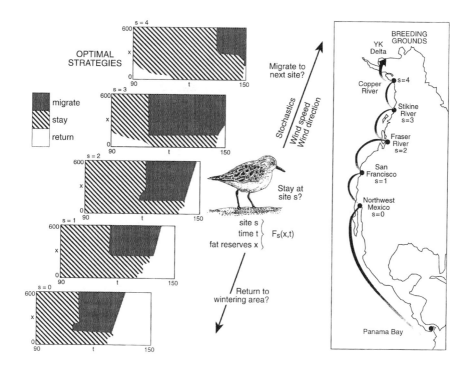

Figure 6.4 Optimal migration strategy for western sandpipers, under favorable wind conditions. The optimal decision $D_s(x,t)$ (migrate, stay, or return) is a function of time t, energy reserves x, and site s. From Clark and Butler (1999).

et al. 1998). Thus, for a bird located at sites 0, 1, or 2, or at sites 3 or 4 after $t = 109$,

$$X(t+1) = X(t) + \Delta e_1 \qquad (6.21)$$

At sites 3 and 4 before $t = 110$,

$$X(t+1) = X(t) - \Delta e_2 \qquad (6.22)$$

where $\Delta e_2 = 10\,\text{kJ/d}$ is the daily metabolic cost at sites 3 and 4 when food is unavailable. It is assumed that birds die of starvation if $X(t+1) < 0$.

The dynamics of $X(t)$ during migratory flights are determined by the distance D between adjacent sites and by the winds encountered en route. Flight speed in still air is assumed constant at $v = 38.5$ km/hr. If the bird encounters wind vector $\overrightarrow{w}$, we assume that it uses the flight vector $\overrightarrow{v}$ (of magnitude v) that compensates for the wind, in the sense that the resultant vector $\overrightarrow{v} + \overrightarrow{w}$ points in the direction of the next site, if that is possible (see fig 6.2). For the basic model, Clark and Butler (1999) assumed that the bird can detect "favorable" wind conditions for the flight, while on the ground. A favorable wind is defined as a $\overrightarrow{w}$ that aids flight, in the sense that $v_g > v$, where v_g

Table 6.1 Probability distributions p_{is} for favorable ground speed $v_g = v_i$ between sites s and $s + 1$

site	ground speed (km/h)					
s	46.3	63.0	81.0	99.0	117.0	153.0
0^a	0.436	0.359	0.205	0	0	0
1	0.436	0.359	0.205	0	0	0
2	0.667	0.167	0.125	0.042	0	0
3	0.298	0.234	0.191	0.213	0.064	0
4	0.457	0.326	0.109	0.065	0.022	0.022

a. Sites 0 and 1 used the same meteorological data.

denotes ground speed, as given by eq. 6.7. Meteorological data indicate that favorable winds occur (at least at some suitable altitude) on 65–80% of spring days along the flyway. (An alternative version of the model, to be discussed later, assumes that western sandpipers cannot detect and utilize favorable winds.)

Although the bird can identify favorable wind conditions, it does not know the actual wind vector $\vec{w}$ it will encounter on the flight. A discrete distribution of favorable ground speeds v_g was estimated from meteorological data: $p_{is} = \Pr(v_g = v_i$ for flight from site s to $s + 1)$. The values of p_{is} used are in table 6.1. This calculation also assumed that western sandpipers can choose the altitude at which winds are most helpful. This point is discussed further later on.

The time required for completing the flight is $t_f = D_s/v_g$ h, where D_s is the distance between sites. For example $D_1 = 1540$ km for the flight from SF to FR, and this is completed in from 19.0 to 33.5 h depending on winds. In the model, these values are rounded up to 1 or 2 d, respectively. These times concur with results of telemetry studies of migrating western sandpipers (Iverson et al. 1996).

The dynamics of fat reserves during migration are modeled as in eq. 6.12. Let $D_{air} = vt_f = vD_s/v_g$ denote the equivalent air distance flown. Then

$$D_{air} = \int_{x_a}^{x} \frac{dx}{J(x)} \tag{6.23}$$

where x and x_a are reserves at departure and arrival, respectively. Equation 6.23 applies provided that the initial reserves are sufficiently large—that is, provided that

$$Y(x) = \int_0^x \frac{dx}{J(x)} \geq D_{air} \tag{6.24}$$

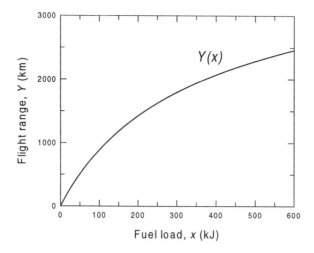

Figure 6.5 Flight range for western sandpipers, basic model ($\alpha = 2.0$).

If this is not the case, we assume that the bird exhausts its reserves during the flight and perishes.

In eq. 6.23, $J(x)$ denotes flight cost in kJ/km. We use the Pennycuick function of eq. 6.11, for which

$$Y(x) = \frac{2}{ab}(1 - \frac{1}{\sqrt{1 + bx}}) \qquad (6.25)$$

To apply eq. 6.25, we need to estimate parameters a and b. To estimate a, we use a formula of Norberg (1996, p. 241), estimated from wind tunnel and other data for 39 species of birds: $P = 57.3M^{0.81}$, where P denotes metabolic cost of flight (watts) and M is body mass (kg). For a western sandpiper carrying a 50% fuel load of 300 kJ, $M = 30$ g $= 0.03$ kg, giving $P = 3.39$ W $= 12.2$ kJ/hr. With airspeed $v = 38.5$ km/hr, this implies that $J(300) = a(1 + 300b)^{3/2} = 12.2/38.5 = 0.317$ kJ/km. Since there are two coefficients a and b, we need a second value for $J(x)$. Lacking any data from which to determine this second value, Clark and Butler introduced a parameter $\alpha = J(600)/J(300)$—that is, α is the increase in flight cost for a bird carrying a full fuel load, relative to a bird with a 50% load. They treated α as a tunable parameter, which was adjusted to obtain the best fit between model predictions and field data for the timing of spring migration. Such parameter tuning was discussed in section 3.2.

The tuned value of α turns out to be $\alpha = 2.0$, implying that $b = 0.0048$/kJ and $2/ab = 5018$ km. This in turn implies that $Y(300) = 1795$ km; the graph of $Y(x)$ is shown in fig 6.5. A clue to the degree of sensitivity to α is provided by noting that changing α by $\pm 25\%$ to $\alpha = 1.5$ or 2.5 changes $Y(300)$ to 1246 and 3409 km, respectively. As will be seen later, the predicted timing of migration is quite sensitive to α.

To calculate the reserves x' that remain on arrival at the next stopover site, we note that by eqs. 6.23–6.25,

$$D_{\text{air}} = Y(x) - Y(x') = \frac{2}{ab}\left(\frac{1}{\sqrt{1+bx'}} - \frac{1}{\sqrt{1+bx}}\right) \tag{6.26}$$

where x denotes the take-off reserve level and D_{air} is total air distance flown. This equation is solved for x', given x and D_{air}.

This completes the description of the dynamics of $X(t)$ for both feeding and migrating. Next, we must specify the terminal fitness. For a bird that arrives on the breeding ground on day t with reserves x, we assume that expected reproduction (of ultimately surviving female offspring) is given by

$$\Phi(x,t) = \Phi_1(x)\Phi_2(t)r_{\text{AK}} \tag{6.27}$$

where r_{AK} is the average number of female offspring that survive to maturity, from a female parent that arrives at the beginning of the breeding season with sufficient fat reserves and where $\Phi_1(x)$ and $\Phi_2(t)$ are penalty factors for reduced reserves and late arrival, respectively. We assume

$$\Phi_1(x) = \min(1, c_1 x) \tag{6.28}$$

$$\Phi_2(t) = \begin{cases} 1 - c_2(t - t_{\text{AK}}) & \text{if } t \geq t_{\text{AK}} \\ 0 & \text{if } t < t_{\text{AK}} \end{cases} \tag{6.29}$$

where c_1, c_2 are constants and t_{AK} denotes the opening of the Alaskan breeding season (table 6.2). The parameter values imply that the low-reserves penalty applies only if reserves are less than 120 kJ, and that the late-arrival penalty increases by 2.5%/d.

Dynamic programming equations

We let $F_s(x,t)$ denote the maximum expected future lifetime reproductive success for a female western sandpiper (before assessing current wind conditions), that is located at site s and carries reserves x at the beginning of day t. We also denote by $f_s(x,w,t)$ fitness after assessing current wind conditions $W(t) = w$, where $w = 0$ or 1 for unfavorable or favorable conditions. Thus,

$$F_s(x,t) = E_w\{f_s(x,w,t)\}$$
$$= p_{\text{fav},s}f_s(x,1,t) + (1 - p_{\text{fav},s})f_s(x,0,t) \tag{6.30}$$

where $p_{\text{fav},s}$ is the probability that winds are favorable at site s.

Then

$$f_s(x,0,t) = \max\{V_{1s}(x,t), V_{2s}(x,t)\}$$
$$f_s(x,1,t) = \max\{V_{1s}(x,t), V_{2s}(x,t), V_{3s}(x,t)\} \tag{6.31}$$

Table 6.2 Western sandpiper migration model parameters

Symbol	Meaning	Base-case values
Global parameters		
t_0, t_1	Initial and terminal dates (Julian)	90,151
x_{max}	Maximum energy reserves (kJ)	600
Δe_1	average energy gain while feeding (kJ/d)	39.0
Δe_2	Metabolic cost if not feeding (kJ/d)	10.0
v	Average flight speed (km/hr)	38.5
c_f	Average flight cost (kJ/hr)	12.2
R_0	Expected future reproduction	1.0
α	Flight cost coefficient	2.0
q_v	Wind optimization ability	1.0
Site-specific parameters $(s = 0, 1, 2, 3, 4)$		
$t_{open}[s]$	Site-opening date	$\{90, 90, 90, 110, 110\}$
$D[s]$	Distance to next site (km)	$\{960, 1540, 1180, 900, 1180\}$
$p_w[s]$	Probability of favorable winds	$\{.65, .65, .80, .78, .78\}$
$m[s]$	Predation risk (/d)	$\{.0005, .002, .003, .003, .003\}$
$m_1[s]$	Return migration mortality risk	$\{.0, .02, .05, .06, .08\}$
$t_1[s]$	Return migration time (d)	$\{0, 3, 8, 12, 15\}$
Breeding-window parameters		
t_{AK}	Opening date, Alaska site	131
r_{AK}	Maximum expected reproduction (females)	0.297 (see text)
c_1	Low energy reserves penalty	$1/(0.2x_{max})$
c_2	Late arrival penalty	0.025

where V_{1s}, V_{2s}, V_{3s} are the fitness values corresponding to (1) remaining at the current site on day t, (2) initiating return migration to the winter site, and (3) initiating migration to the next site $s + 1$. Recall that the bird can opt for migration to the next site only when $w = 1$, that is, winds are favorable. The functions $V_{is}(x, t)$ are obtained as follows. For $i = 1$ (remain on site)

$$V_{1s}(x, t) = [1 - m(s)]F_s(x + \Delta e_{st}, t + 1) \qquad (6.32)$$

where Δe_{st} equals net energy gain (or loss) obtained by feeding for the day on site s, day t, as explained above, and where $m(s)$ is the daily mortality risk from predation at site s. (Note that predation risk is not assumed to be mass-dependent in this model.) For $i = 2$ (return migration),

$$V_{2s}(x, t) = \sigma(s, t)R_0 \qquad (6.33)$$

where $\sigma(s,t)$ is the probability of surviving from day t in the current year to day $t_0 = 90$ next year, for a bird that initiates return migration from site s on day t, and R_0 is the expected future lifetime reproductive success for an adult female located at the winter site on day t_0. Life-history aspects of the model are discussed in greater detail later. Finally, for $i = 3$ (initiate migration),

$$V_{3s}(x,t) = E_{t_f}\{F_{s+1}(x'_{t_f}, t + t_f)\} \tag{6.34}$$

where t_f is the time to complete the migration, as described before, and where x'_{t_f} is energy reserves upon arrival at the next site, as in eq. 6.26. Recall that winds actually encountered during the migration are stochastic, and thus eq. 6.34 includes an expectation relative to the time spent on migration.

Two terminal conditions pertain to the terminal site $s = 5$ (Alaska) and the terminal time $t = T = 151$. At site 5, if $\Phi(x,t)$ is expected current reproduction (eq. 6.27) and σ_5 is the probability of survival from day T this year to day t_0 next year, for a bird located in Alaska, then

$$F_5(x,t) = \Phi(x,t) + \sigma_5 R_0 \tag{6.35}$$

It is assumed that the survival coefficients $\sigma(s,t)$ and σ_5 reflect both the risks of the return migration and of overwinter survival, which are taken to be related to the distance migrated and the time spent on the wintering ground, respectively; see table 6.2.

At other sites, for $t = T$, we set

$$F_s(x,T) = \sigma(s,T)R_0 \tag{6.36}$$

that is, at day T, a bird located at site s does not complete the spring migration, returns to the winter site, and hence obtains fitness only from reproduction in future years.

Parameter values

Several parameter values used in the sandpiper migration model (table 6.2) are either poorly known or completely unknown. In some cases, parameters could be estimated from data for related species, but other values had to be specified more or less arbitrarily (Clark and Butler 1999). Predation risks $m(s)$ at winter and stopover sites s are an important example of unmeasured parameters. The actual values of $m(s)$ turned out to be less important than the relative values of $m(s)$ at the wintering site and at stopover sites. As we discuss later, realistic predictions are obtained if $m(0) < m(s)$ for $s = 1, 2, 3, 4$, but completely unrealistic predictions result if this assumption does not hold.

Predictions from the model

Figure 6.4 shows the decision matrices $D_s(x,t)$ under favorable wind conditions for migrating western sandpipers. For each site s, the model predicts

Table 6.3 Modal predicted peak dates for migrating western sandpipers (100 forward iterations of 20 birds) for (a) basic model, (b) observed data, and (c–e) alternative values of tuned parameters α, q_w, as described in the text

Parameters	San Francisco Bay (SF)	Fraser River Delta (FR)	Stikine River Delta (ST)	Copper River Delta (CR)
(a) Predicted	115	119	122	128-129
(b) Observed	111–118	118–121	123–126	129
(c) $\alpha = 1.5$	111	118	122	130
(d) $\alpha = 2.5$	121	123	124	129
(e) $q_w = 0.8$	108	115	118	130

both the earliest date of onward migration and the minimum energy reserves at which birds will migrate:

Site	ME	SF	FR	ST	CR
Earliest migration date	112	117	121	110	134
Minimum reserves (kJ)	150	270	180	180	330

We do not know how well these predictions would bear up to testing in the field. Some of the regions shown in the decision matrices of fig 6.5 would never actually occur, for an optimal migrant. For example, birds will depart from site 2 only at days $t \geq 121$, so none will arrive at site 3 before day 122. Therefore, all of the decision matrix $D_3(x, t)$ for $t < 122$ is irrelevant. This phenomenon, where parts of the decision matrix are never used, commonly occurs in solving dynamic state variable models and can be slightly confusing until you recognize it.

By running Monte Carlo forward iterations one can generate annual migration schedules of birds using the optimal strategy. The iterations that were performed allowed for variations in the dates of arrival of tropical birds at the initial site in Northern Mexico (site 0). In any given year, all migrants encounter identical wind patterns and follow identical migration patterns unless they have low energy reserves or arrive at site 0 at different times. The iterations thus predict that birds will pass through any site in several waves in response to wind conditions, as in fact they do. Table 6.3 shows the predicted modal dates of maximum population numbers at each stopover site for the basic set of parameter values, as well as for several alternative parameters. The observed mean peak dates are also given. The model clearly performs satisfactorily, as far as predicting peak dates is concerned. In part, this is due to tuning of the flight-cost parameter α. Varying α by $\pm 25\%$ from the base value $\alpha = 2.0$ affects the peak dates somewhat (table 6.3, lines c and d), especially at the lower latitude sites, SF and FR. This result emphasizes the importance of accurately estimating the flight-range function, which is poorly known for most species.

The effects of winds

The bottom line (e) in table 6.3 shows the effect of altering the assumption about how well migrating birds optimize the use of winds. In the basic model, this ability (represented by the multiplier q_w) is assumed perfect: birds in flight locate the altitude where winds are most favorable. In the alternative calculation q_w is reduced to 80%. As a result (table 6.3 e) the birds take longer to fly between stopover sites, and hence it is predicted that they will arrive and depart earlier from the more southern sites. This calculation suggests that taking advantage of favorable winds on migration is an important component in the migratory strategy of small birds such as the western sandpiper (Butler et al. 1997).

We also consider the fitness consequences of assuming that the sandpipers cannot detect and optimize the use of favorable winds. To treat this question, we compare lifetime fitness $F_0(x_0, t_0)$ at site 0, day t_0, for the basic model with the same value for an alternative version of the model in which the birds are assumed not to be able to tell whether winds are favorable, nor to be able to optimize winds assistance while flying. (This alternative assumption requires modification of the dynamic programming equations.) The prediction is that this inability to utilize wind assistance would reduce the birds' fitness by about 90%, suggesting that the western sandpiper must in fact possess these abilities to a strong degree.

Predation risk differential

To evaluate the importance of the assumption that predation risks are lower at the winter site than at stopover sites, we ran the model under the assumption that predation risks of .001/d were the same at all sites. This completely changed the model predictions: now all birds migrate immediately (given favorable winds) to the Fraser River delta, where they await the opening of feeding sites further north. This shows that the assumption of differential predation risk is sufficient to explain the observed pattern of migration for western sandpipers. Whether the assumption is valid awaits confirmation.

Life-history consistency

As noted earlier, reproductive value R_0 is both an input to and an output from the migration model. Moreover, the value of R_0 must equal 1.0 for a stable population (see chapter 12), which we assume to be the case for western sandpipers. Along with other wader species, western sandpipers were hunted prior to the Migratory Bird Treaty of 1916. Presumably the population has recovered by now, however; we know of no information to suggest that the current population is either increasing or decreasing.

Of course it is unlikely in the first instance that the calculated value of $F_0(x_0, t_0)$ will equal 1.0, even approximately. If not, the conclusion is that the model is logically inconsistent. Several model parameter values are not accurately known; altering the value of a parameter changes the value of $F_0(x_0, t_0)$. We chose to adjust the parameter r_{AK}, which includes juvenile survival, since this has a direct influence on the fitness of parents. Juvenile survival rates for western sandpipers are also completely unknown and doubtlessly difficult to measure. The value $r_{AK} = 0.297$ given in table 6.1 is the result of adjusting r_{AK} to achieve the consistency requirement that $F_0(x_0, t_0) = R_0 = 1.0$. Since most western sandpipers lay four eggs, two of which produce female chicks, this corresponds to a juvenile survival rate of $\frac{1}{2} r_{AK} = 14.8\%$.

Discussion

The western sandpiper migration model has revealed important gaps in our understanding of bird migration. Paramount is the influence of predation risk. Predators such as gulls and falcons are known to take large numbers of migrating birds (Alerstam 1990, p. 343ff.), and it is reasonable to suppose that migration strategies have evolved to take account of these risks. Our knowledge of the detailed movements of birds, both predators and prey, is rapidly increasing as a consequence of recently available tracking technology, so that testing the predictions of models, like the one presented here, may soon become a reality.

Our model also suggests that the ability to utilize favorable winds nearly optimally may be an important component of fitness (see also Butler et al. 1997). Although the model does not include foraging strategies at refueling stopovers or mass-dependent predation risks, these may also be related to the birds' ability to optimize the use of winds, that is, using favorable winds reduces fuel consumption in migration and hence reduces the time needed to refuel, as well as the optimal departure load, both of which may reduce predation risk (Weber et al. 1998).

The view is sometimes expressed that models should not be attempted and certainly not published, unless sufficient data are available from which to estimate the necessary parameter values. It should be clear by now that we emphatically disagree with this philosophy. On the contrary, we believe that good models can help immeasurably in identifying important gaps in the data or important experiments that need to be carried out. This is indeed a perfectly normal scientific procedure: models (or theories) suggest measurements that can support or refute these models. In contrast, collecting reams of data with no basis in theory is often just a waste of time and effort.

The migration model encompasses a relatively brief, but important, segment of a western sandpiper's yearly schedule, excluding breeding behavior, return migration, molting, and winter fattening strategies. In principle, all these components could be included in a single, grand, full-life model. What further insights into shorebird biology would result from such a model will not be obvious until the attempt is made.

The principal predictions of the western sandpiper model are summarized here.

1. Timing of spring migration:
   (a) Earliest and latest dates of departure from stopover sites
   (b) Mean dates of peak abundance at each stopover site

2. Minimum fuel loads at departure
3. Importance of winds, and of differential predation risks, as factors affecting the evolution of migration strategies
4. Juvenile survival rate (an inverse prediction; see section 3.2)
5. Possible impact of environmental change on the fitness of western sandpipers (see chapter 8)

7

Human Behavioral Ecology

Techniques and concepts used in chapter 7

Forward iteration (7.1, 7.2)
Model testing (7.1, 7.2)
Sensitivity testing (7.1, 7.2)
Cultural anthropology (7.2)
Multidimensional state vectors (7.2)

In this chapter we discuss how dynamic state variable models can be applied to understand problems of human behavioral ecology. Currently some controversy exists about whether any kind of "evolutionary" argument may apply to humans, because our ecological situation is now so different from that in which we evolved (Borgerhoff Mulder and Judge 1993; Crawford 1993; Emlen 1995; Mace 1995, 1996). We avoid this controversy by focusing on two systems in which the assumptions that underlie the use of dynamic state variable models are likely to be justified.

7.1 Dynamic discarding in fisheries

Gillis et al. (1995a) developed a dynamic state variable model for high-grading in the sablefish fishery in Oregon. High-grading occurs when less valuable fish are discarded in anticipation that more valuable fish will be captured later in a fishing trip. Fish will be discarded (and typically killed incidentally in the process) because the fishing vessel's hold has finite capacity; saving low-value fish now means that higher value fish cannot be retained at a later time.

Thus discarding is a source of incidental mortality for low-value fish. By not taking account of this source of mortality, one may miss key features of the population dynamics of some species or misrepresent features of the ecosystem, with dire consequences for both the particular stocks and the ecosystem as a whole (Alverson et al. 1994; Vincent and Hall 1996; Roberts 1997). However, discarding behavior will usually not be observable by the management authority, and failure to account for it will affect the quality of fishery data. Consequently, a model of discarding behavior allows one to infer the characteristics of the actual fishing mortality from the delivered catch.

Gillis et al. (1995a) wanted to understand how the likelihood of high-grading would be affected by time within the fishing trip, the availability of fish, trip limits, and the chance that the trip might end prematurely. Gillis et al. (1995b) applied the ideas to detailed management questions. First we summarize their model (using different notation) and then discuss the particular application to the Oregon fishery for sablefish (*Anaplopoma* spp). Time is measured by the number of hauls that the vessel makes, and the maximum $T = 15$ hauls. The maximum capacity, x_{max}, of a vessel is about 6800 kg of fish, which Gillis et al. discretized into about 30 levels in units of 226.8 kg (500 lb).

Gillis et al. considered that fish could be characterized by one of three categories, indexed by i, from most valuable ($i = 1$) to least valuable ($i = 3$). The economic value of each category was estimated from price data available for 23 trips; the average values were \$1.49/kg, \$0.99/kg, and \$0.61/kg for the three categories, respectively. Thus, we set

$$X_i(t) = \text{weight of fish of value category } i \text{ stored} \qquad (7.1)$$
$$\text{on board at the start of haul } t$$

and have the capacity constraint that

$$\sum_{i=1}^{3} X_i(t) \leq x_{max} \qquad (7.2)$$

Gillis et al. also discretized the weight of a haul with intervals of 500 lb to a maximum of 2,000 lb. Thus there are five categories of weight, and we denote the weight of fish in haul category j by w_j, with $j = 1$ to 5.

Since hauls have probabilistic combinations of weights of fish, Gillis et al. estimated

$$p(w_1, w_2, w_3) = \text{probability that a single haul has weight } w_1$$
$$\text{fish of the highest value, weight } w_2 \text{ fish of} \qquad (7.3)$$
$$\text{the moderate value, and weight } w_3 \text{ fish of}$$
$$\text{the lowest value}$$

To estimate $p(w_1, w_2, w_3)$, Gillis et al. matched the data from 50 hauls with 1,000 simulated hauls, using linear and logistic regression (to take into account the situations in which no fish of a certain value category were found in a haul). This procedure generated $p(w_1, w_2, w_3)$ that matched the variability and correlation structure of the original data. Once the fish are captured in the haul, the discard decision consists of either keeping all of a certain value category or discarding it. We can summarize these decisions by a set of three values $\{d_1, d_2, d_3\}$, in which each entry is either zero or one with the interpretation that if $d_i = 1$, then fish of value category i are kept and if $d_i = 0$, fish of value category i are discarded.

Now the state dynamics are that if the decision triple $\{d_1, d_2, d_3\}$ is applied then with probability $p(w_1, w_2, w_3)$,

$$
\begin{aligned}
X_1(t+1) &= X_1(t) + d_1 w_1 \\
X_2(t+1) &= X_2(t) + d_2 w_2 \\
X_3(t+1) &= X_3(t) + d_3 w_3
\end{aligned}
\tag{7.4}
$$

At the end of a trip, the value of the catch is determined by the mixture of the three types of fish, unless the total catch, summed over the three value categories, exceeds the regulatory quota Q in which case the value is zero. This extreme assumption and some alternatives are discussed by Gillis et al. (1995a).

If the price of fish in category i is v_i dollars/500 lb, the value of a trip when the accumulated catch is $\{x_1, x_2, x_2\}$ is given by

$$
\Phi(x_1, x_2, x_3) = \begin{cases} \sum_{i=1}^{3} v_i x_i & \text{if } \sum_{i=1}^{3} x_i \leq Q \\ 0 & \text{if } \sum_{i=1}^{3} x_i > Q \end{cases}
\tag{7.5}
$$

Now we introduce the "fitness function" (value function might be a better term)

$$
\begin{aligned}
F(x_1, x_2, x_3, t) = \ &\text{maximum expected value of a trip when} \\
&\text{the catches at the start of period } t \text{ are } x_i \\
&\text{of value category } i = 1, 2, 3
\end{aligned}
\tag{7.6}
$$

Gillis et al. assume that there is a probability ρ that a trip will end prematurely and that discarding decisions have to be made before the precise weight of fish from each value category from a haul is known—that is, the discard decision is made before the haul is sorted, and the fish selected for discarding are thrown away as the haul is sorted. With these additional assumptions, the end condition is given by

$$
F(x_1, x_2, x_3, T) = \Phi(x_1, x_2, x_3)
\tag{7.7}
$$

and the dynamic programming equation for $F(x_1, x_2, x_3, t)$ is

$$
\begin{aligned}
F(x_1, &x_2, x_3, t) \\
&= (1 - \rho) \max_d \Sigma p(w_1, w_2, w_3) F(x_1 + w_1 d_1, x_2 + w_2 d_2, x_3 + w_3 d_3, t+1) \\
&\quad + \rho \Phi(x_1, x_2, x_3)
\end{aligned}
\tag{7.8}
$$

Table 7.1. The effects of trip quotas on discarding (other parameters at base case level)[a]

	Trip quota (lb)		
	6,000	8,000	12,000
Medium-value fish discarded (kg)	474.9	331.6	235.9
Low-value fish discarded (kg)	3247.3	2592.9	1437.4
Medium-value fish discarded (%)	19.2	13.2	9.0
Low-value fish discarded (%)	87.5	69.2	36.6
Average length of trip (no. of hauls)	11.5	11.7	12.4

a. From Gillis et al. (1995a).

During the three years of the field study, the management agencies used three different trip quotas: 6,000 lb, 8,000 lb and 12,000 lb, so Gillis et al. investigated all three. (There were also other regulations such as a 5,000-lb trip quota on extra small fish that were not included in the model.) They estimated ρ from the observed occurrence of premature trip terminations due to injury, gear damage, or bad weather. This was lowest in October and November (1 in 140 hauls ended prematurely) and greatest in December–January (13 of 114 hauls ended prematurely), so they examined cases in which $\rho = 0$, .04, .08, or .12. Gillis et al. defined a "base case" consisting of a 12,000-lb quota, average availability of fish, and no risk of premature termination of the trip.

After estimating parameters, Gillis et al. simulated 500 trips using forward iteration. It was not possible to compare the predictions of the model directly with field data because discard amounts are not usually recorded during the fishing trips. A fraction of the trips had observers, so that at least qualitative comparison with some of the data was possible.

There is no reason to discard the highest value fish (unless the hold is already full), but the amount of medium and low value fish discarded can depend on the trip quota (table 7.1), the availability of fish (table 7.2), and the risk that the trip will end prematurely (table 7.3); these results are averaged over the 500 simulated fishing trips. The least valuable category is discarded the most, both in weight and percentage of catch but some of the trends in these tables are not completely intuitive (and it is worthwhile to spend time thinking about them). Decreasing the trip quota increases the predicted amount of discarding and decreases the predicted length of a trip (table 7.1). It is predicted that an increase in the overall availability of fish will increase the discards of the lowest value fish, both in weight and percentage (table 7.2) but is likely to have little effect on the medium value fish. Furthermore, Gillis et al. predicted that the length of a trip will increase with increased availability of fish. Gillis et al. reported, but did not show, that the frequency distributions of the discarding were highly skewed: for example, with the largest trip quota and other base case parameters, on average, 235.9 kg of medium value fish were

Table 7.2. Effect of the availability of fish on discarding (other parameters at base case level)[a]

	Fish availability[b]		
	Low	Medium	High
Medium-value fish discarded (kg)	284.4	235.9	220.0
Low-value fish discarded (kg)	923.5	1437.4	2600.4
Medium-value fish discarded (%)	9.6	9.0	8.0
Low-value fish discarded (%)	27.2	36.6	55.2
Average length of trip (no. of hauls)	11.4	12.4	12.8

a. Gillis et al. (1995a).
b. For low fish availability, the elements of $p(w_1, w_2, w_3)$ were multiplied by 0.5, and for high availability by 2.0. Medium availability is the base case of the model.

Table 7.3. Effects of premature termination on discarding (other parameters at base-case level)[a]

	Risk of premature termination, ρ			
	0.0	0.04	0.08	0.12
Medium-value fish discarded (kg)	235.9	126.1	59.4	40.4
Low-value fish discarded (kg)	1437.4	781.1	461.3	29.8
Medium-value fish discarded (%)	9.0	6.3	3.7	3.0
Low-value fish discarded (%)	36.6	25.7	18.9	14.9
Average length of trip (no. of hauls)	12.4	9.2	7.1	5.6

a. From Gillis et al. (1995a).

discarded. However, the range was 0 kg to 4.3 metric tons (1 mt = 1,000 kg), and the median value was < 1 kg. Thus, for many trips, there was little or no discarding of the medium value fish, but when discarding occurred, large amounts of medium value fish were discarded. Such highly aggregated frequency distributions are common in cases of incidental catch (Mangel 1993; Hilborn and Mangel 1997, chapter 4).

In general, it is predicted that the likelihood of high-grading will increase as the trip progresses (figs. 7.1–7.3); this is consistent with observer information that Gillis et al. reported. In these figures, all parameters are kept at their base-case values, except for the one parameter that is varied as indicated.

Then Gillis et al. compared the predictions of the model to the limited observer data for the Oregon sablefish fishery. Thirty-six trips were observed when no quota was in effect during fishing. Although a proportion of the catch was discarded, there was no relationship between the proportion of catch

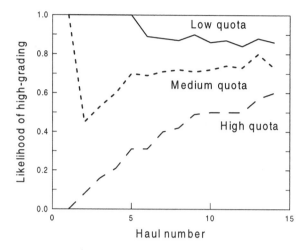

Figure 7.1 The fraction of hauls with discarding as a function of haul number within a trip, and trip quota.

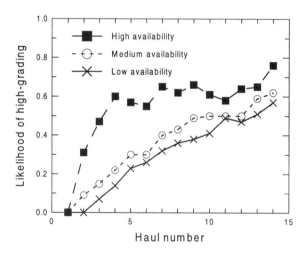

Figure 7.2 The fraction of hauls with discarding as a function of haul number and fish availability.

discarded and the availability of fish ($p > .98$). On the other hand, in the 12 trips that were observed when a quota was in effect during fishing, there was a statistically significant relationship ($p < .03$) between the proportion of fish discarded and the availability of fish. Furthermore, Gillis et al. found that the tendency to discard increased as the trip length increased. These results are consistent with the predictions of the dynamic state variable model. Gillis et al. (1995b) investigated further management implications of the results from the model.

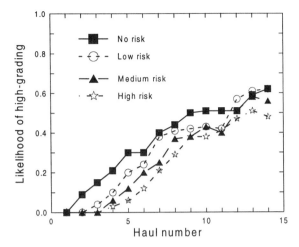

Figure 7.3 The fraction of hauls with discarding as a function of haul number and the chance that the trip will end prematurely.

In summary, the model leads to the predictions that high-grading will be more common toward the end of a fishing trip, when there is an increase in the overall availability of fish, a decrease in the trip quota, or a decrease in the risk of a premature termination of the trip. None of these results is particularly surprising. However, the model also showed that an increase in the availability of fish may have no effect on the discarding of some value classes of fish and that high-grading may be more likely early in a trip, if there is a chance of filling the quota with the most valuable category (e.g., when the availability of fish is high and the quota is low). Furthermore, the model can be used to estimate the incidental mortality on the different value classes caused by high-grading and to help set regulations. For example, larger quotas will reduce the likelihood of high-grading but will also increase overall fishing mortality. Models such as that described in this section can be used to predict the consequences of different regulatory schemes. For example, the models can be used to minimize total mortality (fishing + discarding) given a catch target consistent with a sustainable fishery.

Other applications of dynamic state variable models to resource management are developed by Hart (1997); Milner-Gulland and Leader-Williams (1992); and Leader-Williams and Milner-Gulland (1993). More general applications of these methods in economics fall into the category of investment under uncertainty; Dixit and Pindyck (1994) is a good starting place.

7.2 Optimal reproductive strategies in !kung Bushmen females

Anthropologists who study !kung (i.e. Bushmen) reproductive behavior have noted that mothers space their children quite far apart and typical interbirth

Table 7.4. Survival rates (to age 10 years) for children, as a function of their mothers' IBIs and expected lifetime reproduction (ELR) of the mothers[a]

IBI (months)	N	Survival rate	ELR[b]
24 (18–29)	6	16.7%	1.67
36 (30–41)	23	39.1%	2.35
48 (42–53)	29	55.2%	2.76
60 (54–65)	16	56.2%	2.25
72 (66–77)	6	100.0%	3.00
84 (78–89)	7	100.0%	2.00
96 (90–101)	2	100.0%	2.00
108 (102+)	7	42.8%	2.00

a. From Blurton Jones 1986.
b. This column differs from the reference in allowing only an integral number of offspring per mother.

intervals (IBIs) range from three to five years (Lee 1972, 1979; Howell 1979). At first glance, these long IBIs might seem to indicate a failure to maximize lifetime reproductive success. Among explanations for this behavior are (1) that children are not highly valued (but this contradicts !kung stated values), (2) that women deliberately limit their reproduction to control population size or to maintain a leisurely life style, and (3) that the work-load for women who use shorter IBIs is simply too great to bear (Lee 1972; Blurton Jones 1986).

Blurton Jones and Sibley (1978) expanded on the third of these explanations by computing a woman's total average backload for different IBIs, assuming that the women provide approximately 60% of the family food in the form of mongongo nuts. In the dry season, these nuts must be retrieved from a distance of several kilometers; making a trip every three days necessitates a load of more than 9 kg of nuts. Besides this, the mother carries all of her children up to age 5 years, in potentially lethal climatic conditions. Blurton Jones and Sibley suggested that, under these circumstances, the !kung women are spacing births to maximize expected lifetime reproduction.

The data given in table 7.4, derived by Blurton Jones (1986) from field data of Howell (1979), support this hypothesis. Here, expected lifetime reproduction (ELR) is calculated as $S(I) \cdot \text{int}(L/I)$, where I is the interbirth interval, $S(I)$ is the child survival rate as listed in the table, and $L = 20$ is the assumed reproductive life-span of an individual. The expression $\text{int}(L/I)$ represents the integer part of L/I; for example, with $I = 6$, $\text{int}(L/I) = 3$ because only three children (not 3.33) can be produced. According to this calculation, the optimal IBI is six years (72 months), but most women choose IBIs from 3 to 5 yr. As pointed out by Blurton Jones (1986), the data for IBIs of 6–8 yr are rather sparse. If even a single death occurred in the 6-yr group, its advantage would disappear.

Table 7.5. Backload data[a]

Age	Child's body weight (kg)[b]	Food weight (kg)	Total backload per child (kg)
0	4.8	1.5	6.4
1	8.1	2.5	10.6
2	8.7	3.5	12.2
3	6.8	3.8	10.6
4	1.2	3.2	4.4
5+	0.0	3.4	3.4

a. From Anderies 1996, table 1.
b. Includes a factor for the proportion of time the child needs to be carried.

Anderies (1996) developed a dynamic state variable model of birth spacing in !kung women, based on the earlier work. In this model, basic time periods are defined as one year, and reproductive decisions are made on this time-scale. The time t is taken as the woman's age, $0 \leq t \leq T$. The state variable $\overrightarrow{X}(t)$ is a vector $(X(t) = (X_0(t), X_1(t), \ldots, X_5(t))$ that has 6 components:[1]

$$X_i(t) = \text{children of age } i \quad i = 0 \text{ to } 4$$
$$X_5(t) = \text{children of age} \geq 5$$

Thus $X_i(t) = 0$ or 1 (ignoring twins, etc.) for $i = 0$ to 4, and $0 \leq X_5(t) \leq X_{5\,\text{max}}$. We set $X_{5\,\text{max}} = 10$ on the grounds that children more than age 15 are largely independent and have no effect on their mother's current reproductive decisions.

The decision $D(\overrightarrow{x}, t)$ refers to the woman's decision to conceive, given her current age t and state $\overrightarrow{X}(t) = \overrightarrow{x}$. Here $D = 0$ means do not conceive, and $D = 1$ means conceive.

The woman's backload B depends in two ways on her current family state $\overrightarrow{X}(t)$. First, young children must be carried for all or part of each foraging trip, depending on their ages. Second, food needs to be retrieved to feed all the children. The total backload is assumed to be given by

$$B(\overrightarrow{X}) = 9.2 + \sum_{i=0}^{5} w_i x_i \text{ kg} \tag{7.9}$$

where the w_i are the requirements of children of age i (table 7.5). Here the quantity 9.2 kg represents three-day food requirements for herself, her

---

[1] We have modified Anderies's (1996) model in several respects to simplify the presentation.

husband, and one dependent relative. Foraging trips are made every three days.

Backload $B = B(\overrightarrow{x})$ determines the probability that mother forages successfully during the year, and this also depends on her age t:

$$p_f(\overrightarrow{x}, t) = \Pr\,(\text{mother forages successfully} \mid \overrightarrow{X}(t) = \overrightarrow{x}) \qquad (7.10)$$

Anderies assumed that $p_f(\overrightarrow{x}, t)$ decreases with increasing backload and with increasing age of the mother (Anderies 1996, fig. 2).

The mother's annual survival probability p_s could also depend on her current backload $B(\overrightarrow{X}(t))$, but lacking the necessary data for estimating this relationship, Anderies assumed only age dependence:

$$p_s(t) = \Pr\,(\text{mother survives from age } t \text{ to } t + 1) \qquad (7.11)$$

Neither $p_f(\overrightarrow{x}, t)$ nor $p_s(t)$ depends on the current decision $D(\overrightarrow{x}, t)$.

If the mother forages successfully, her children survive with age-specific probabilities σ_i ($i = 0$ to 5). If she fails to forage successfully, it is assumed that all children less than age 5 die of malnutrition, but she and her other children survive the year. Finally, if the mother dies, all of her young children also die, but older children may survive to reach maturity, with a probability σ_i that depends on their ages.

This information suffices to specify the state dynamics. For example, consider the case $\overrightarrow{X}(t) = \overrightarrow{x} = (0, 1, 0, 0, 1, 3)$ that is, the mother has a 1- and a 4-year old, plus three older children. If she does not conceive, forages successfully, and survives, then

$$\overrightarrow{X}(t + 1) = \overrightarrow{x}' = \begin{cases} (0, 0, 1, 0, 0, 4) & \text{with probability } \sigma_2 \sigma_4 \sigma_5^3 \\ (0, 0, 0, 0, 0, 4) & \text{with probability } (1 - \sigma_2)\sigma_4 \sigma_5^3 \end{cases} \qquad (7.12)$$

and so on (how many different outcomes are possible?). If she conceives, then $\overrightarrow{x}'$ is the same, except that $x_0' = 1$. If she fails to forage successfully, then $\overrightarrow{x}' = (0, 0, 0, 0, 0, 3)$ with probability σ_5^3, and so on. Anderies did not include child mortality other than from malnutrition in his dynamic programming formulation, although he did consider the possibility in forward simulations. Without redoing his calculations—a considerable chore!—it is not clear what effect this simplification would have on the predicted optimal decision matrix $D^*(\overrightarrow{x}, t)$.

Fitness $F(\overrightarrow{x}, t)$ is defined as maximum expected total reproduction from age t to T, given that $\overrightarrow{X}(t) = \overrightarrow{x}$. Since the mother's actions affect her living children and also produce newborns, reproductive success is specified as the total expected number of children that survive to sexual maturity. Note in particular, that including the states of the children in the mother's state variable $\overrightarrow{X}$ allows us to treat decisions pertaining to long-term parental care. Terminal fitness is given by

$$F(\overrightarrow{x}, T) = \Sigma \delta_i x_i \tag{7.13}$$

where $\delta_i = \Pr\{\text{child of age class } i \text{ survives to maturity without direct parental care}\}$. Here, all children of age class 5 are considered equivalent, and δ_5 is the average survival probability of children of age ≥ 5.

The dynamic programming equation has the form

$$F(\overrightarrow{x}, t) = \max\{V_0(\overrightarrow{x}, t), V_1(\overrightarrow{x}, t)\} \tag{7.14}$$

where the subscripts 0 or 1 refer to the reproduction decision D, that is, skip a year, or conceive. For example

$$
\begin{aligned}
V_0(\overrightarrow{x}, t) \\
= p_s(t)\{p_f(\overrightarrow{x}, t)EF(\overrightarrow{x}', t+1) + (1 - p_f(\overrightarrow{x}, t))EF(\overrightarrow{x}'', t+1)\} \\
+ (1 - p_s(t))\Sigma \delta_i x_i
\end{aligned}
\tag{7.15}
$$

A similar formula holds for $V_1(\overrightarrow{x}, t)$. Here $\overrightarrow{x}'$ and $\overrightarrow{x}''$ denote the state $\overrightarrow{x}(t+1)$ when the woman forages successfully or not, respectively. These are random variables, as explained in the discussion of eq. 7.12, and the symbol E denotes expectation over these random variables.

As noted, the functions $p_s(t)$ and $B(\overrightarrow{x})$ can be estimated from available data, but the foraging success probability $p_f(\overrightarrow{x}, t)$ cannot be estimated from data and must be determined hypothetically. Sensitivity tests can be used to evaluate the effect of alternative assumptions about $p_f(\overrightarrow{x}, t)$.

Predictions of the model

Using forward iteration, Anderies (1996) obtained optimal reproduction histories for !kung women, based on this model. Typical examples shown in fig. 7.4 correspond to two different specifications of $p_f(\overrightarrow{x}, t)$: case 1 (maximum backload 40 kg), and case 2 (maximum backload 50 kg). IBIs decrease with increasing backload capacity (3.33 yr and 3.0 yr, respectively), supporting the suggestion that backload is a critical determinant of birth spacing. In both cases, IBIs increase with age; older women have growing families to feed and cannot increase their families beyond backload capacity levels. This common-sense prediction agrees with field data (Blurton Jones 1987).

The dynamic state variable model facilitates the consideration of questions other than optimal IBIs. For example, Anderies (1996) considers the question of menopause, that is, termination of reproduction prior to normal life-span. In modeling menopause as a possible adaptation, Anderies assumed that conception remained possible throughout the mother's life. His model predicts ages of last birth ranging from 45–60 yr, depending on parameter

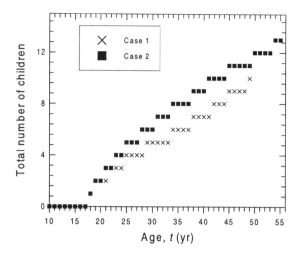

Figure 7.4 Optimal reproductive histories for !kung women (ignoring possible child mortality): case 1, maximum backload 40 kg; case 2, maximum backload 50 kg. From Anderies (1996).

values. In general, increased survival rates for children imply lower ages of menopause, which makes good sense. The hypothesis that human menopause is an adaptation to the long period of adolescence has been widely discussed (e.g., Hill and Hurtado 1991; Mayer 1982).

Anderies also considered an important cultural aspect of !kung life, namely, the dependence of children's survival on their mother's provisioning. The basic model predicted that a woman would not have more than two children under age 8 at any time. This contrasts with other cultures, in which several children are often produced in rapid succession. The !kung model makes similar predictions if the assumption of strict dependence of children's survival on their mother's provisioning is altered, so that children will be taken care of regardless of the mother's success at food retrieval.

To summarize, Anderies' dynamic state variable model of !kung reproductive behavior was successful in predicting dynamic interbirth intervals that agree with field data. It also facilitated sensitivity studies and threw light on cultural aspects of !kung society.

8

Conservation Biology

Techniques and concepts used in chapter 8

Environmental impact assessment (8.1)
Computing binomial coefficients (8.2, 8.3)
Metapopulation dynamics (8.3)
Marine reserves (8.4)

8.1 Assessing the fitness consequences of environmental change

A dynamic state variable model yields numerical predictions of fitness—that is, expected life-time reproductive success, for organisms that use optimal decision strategies under the assumptions of the model. If the parameters of the model are altered, these optimal strategies and fitness values will also change. In previous chapters, we have considered this relationship between parameters and model predictions in terms of sensitivity testing of the model. The same method can also be used for environmental impact assessment. Thus a potential change in the organism's environment may be reflected in changes in the model's parameters. By rerunning the calculations with the new parameter values, we obtain new fitness values. Then comparing the original and altered fitness values provides a prediction of the relative effect of the environmental change. (Policy makers might find this result more comprehensible if it were phrased as a certain percentage change in the species' net productivity, rather than as a change in fitness.)

Note, however, that this calculation tacitly assumes that the organism in question alters its behavior to achieve the new optimum. Unless the modified environmental conditions are within the range normally encountered by the organism, this change can evolve only by natural selection. To calculate the

Table 8.1 Percent reduction in fitness resulting from a 50% decline in food availability at separate stopover sites and from the same decline at all sites simultaneously[a]

Site	No adaptive response (%)	New optimum (%)
San Francisco	0.2	0.2
Fraser River	4	2
Stikine River	5	2
Copper River	20	8
All sites	35	17

a. From Clark and Butler 1999.

short-term (nonevolutionary) change in fitness, we can modify the code for the model, using the altered parameter values and assuming that the organism continues to use the originally optimal decision strategy $D(x, t)$—which is now usually no longer optimal (see section 2.6).

The two approaches just described can be used to assess short- or long-term evolutionary changes in fitness. The method cannot predict the rate of transition between these short- and long-term responses; this would require a full-scale genetic model. The method also cannot assess changes in population abundance, which would require understanding density-dependent mechanisms operating at each stage of the organism's life cycle and how behavior interacts with these mechanisms (Sutherland 1996). There are probably few, if any, wild populations for which this degree of information is currently available. The dynamic state variable approach, though perhaps limited, nevertheless might be useful in comparing the potential impacts of different likely changes to the environment. For example, the prediction that one type of change would reduce a given species' productivity by 5%, whereas another would reduce it by 50%, might be of considerable interest.

As an illustration, we consider again the western sandpiper migration model described in chapter 6. Table 8.1 shows the effects of a 50% decline in food availability at separate refueling stopover sites and also at all four sites simultaneously. Environmental events that might affect individual sites include industrial development, pollution, and oil spills. A rise in sea level that results from global warming could affect the food availability at all the sites.

Note that the predicted effect of food reduction at any given site is fairly minor. The effect of simultaneous change on all sites is more severe, especially in the short term. It is easy to understand why the model predicts these results. First, taking advantage of favorable winds is more important for western sandpipers than rapid energy gain at any given stopover. Obtaining food at a given stopover site is of relatively minor importance. However, if the food supply is reduced at all the refueling sites, the migrants must spend

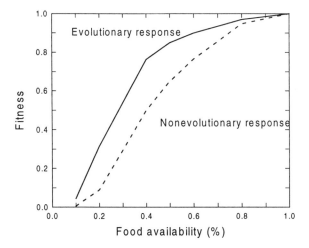

Figure 8.1 Fitness $F_0(x_{max}, t_0)$ as a function of food supply at all sites simultaneously for the short-term (nonevolutionary) and long-term (evolutionary) behavioral response of western sandpiper migration strategies. From Clark and Butler (1999).

more time foraging at each site. If they continue to use the original decision strategy (the short-term case), this delays their average arrival date in Alaska and reduces their reproductive success. Predation risk on stopover sites is also increased. Over the long run, on the other hand, the birds evolve a new strategy and leave the winter site at an earlier date with higher levels of energy reserves. This new strategy significantly increases fitness relative to the original strategy.

To what extent can these predictions be applied to other species of migratory birds? Clearly this question cannot be answered quantitatively without developing specific dynamic models for the species in question. However, the predictions from one model can at least suggest phenomena that might pertain to other species.

For example, the question arises whether progressive environmental changes are likely to have a threshold effect on migratory bird populations. Progressive deforestation along the migratory route of a song bird, for example, might initially have minimal impact, but beyond some critical level, could severely reduce the bird's chances of completing the migration successfully. Dynamic state variable models provide an approach to addressing such questions.

Figure 8.1 shows the fitness values $F_0(x_{max}, t_0)$ for the western sandpiper model, as a function of the percent decline in food availability over all sites, for both short- and long-term responses. These curves do not indicate any sharp threshold for food availability, but fitness does decline rapidly when the food supply falls much below 50% of the estimated current level.

8.2 Using fire rotation schemes to maximize the persistence of midsuccessional species

Most of the examples discussed in this book deal with applications of the methods of stochastic dynamic programming to problems of individual behavior. However, there is a long and rich history of application of these methods to problems of resource management (see Watt 1968; Walters and Hilborn 1978; Mangel 1985; Walters 1986; Clark 1990; Moore 1990; B.K. Williams 1996 for various reviews). Recently, Possingham and colleagues argued that dynamic, state-dependent thinking is essential for effective conservation (Possingham 1997) and gave examples that include designing a network of reserves to maximize protection of biodiversity (Possingham et al. 1993), managing a metapopulation (Possingham 1996), and determining fire rotation schemes (Possingham and Tuck 1997). We discuss the latter two examples, beginning with fire rotation, in a form slightly different from that used by Possingham and Tuck (1997).

Many threatened or endangered species prefer early or midsuccessional habitats (Johnson 1992; Friend 1993; Baker and Whelan 1994; Whelan 1995; Pyke et al. 1995)—that is, as the time since the last fire increases, the quality of habitat improves, peaks, and then declines. We will assume that habitat quality is a determinant in the population dynamics of the endangered species.

Following Possingham and Tuck (1997), we let $q(t)$ denote the relative quality of habitat t years after a fire and set (fig. 8.2)

$$
q(t) = \begin{cases}
t/t_1 & \text{if } 0 \le t \le t_1 \\
1 & \text{if } t_1 < t \le t_2 \\
1 - (1 - q_0)(t - t_2)/(t_3 - t_2) & \text{if } t_2 < t \le t_3 \\
q_0 & \text{if } t > t_3
\end{cases} \tag{8.1}
$$

where habitat quality peaks at $t = t_1$, begins to decline at $t = t_2$, and reaches a steady level q_0 at $t = t_3$.

Population dynamics in a successional environment

We use a population model slightly more complicated than that used by Possingham and Tuck (1997). Because threatened or endangered species are usually at low population sizes, we can ignore most kinds of density dependence when studying the population dynamics. Clearly, this is not true for Allee effects, in which reproduction or survival are reduced at low population density (Clark 1990), but we will not consider those effects here. We assume that the population processes are structured according to (1) reproduction, (2) natural survival of juveniles and adults, and (3) additional mortality caused by the fire. We also assume, for simplicity, that there is a 1:1 sex ratio and that juveniles who survive become reproductive in the next season.

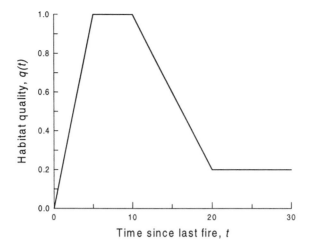

Figure 8.2 The relationship proposed by Possingham and Tuck (1997) between habitat quality and time since the last fire: see eq. 8.1 with $t_1 = 5$, $t_2 = 10$ and $t_3 = 20$.

Thus we let

$$N(t) = \text{number of reproductive females at the start of year } t$$
$$R(t) = \text{reproduction in year } t \tag{8.2}$$

Reproduction is determined by habitat quality $q(t)$ and clutch or litter size r per reproductive adult according to

$$R(t) = rq(t)N(t) \tag{8.3}$$

In the absence of fire, we assume that an adult survives with probability s_a and that a juvenile survives with probability s_j. Then, the number of reproductive females at the start of year $t + 1$ will be the sum of two binomially distributed random variables: one characterizes the number of adults who survive, N_a, given $N(t)$, and one characterizes the number of juveniles N_j who survive, given $R(t)$: These are

$$\Pr\{N_a = n | N(t) \text{ adults}\} = \binom{N(t)}{n} s_a^n (1 - s_a)^{N(t)-n}, \ 0 \le n \le N(t)$$

$$\Pr\{N_j = m | R(t) \text{ juveniles}\} = \binom{R(t)}{m} s_j^m (1 - s_j)^{R(t)-m}, \ 0 \le m \le R(t)$$

$$\tag{8.4}$$

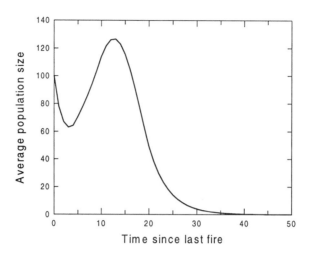

Figure 8.3 Trajectory of the average population size given by eq. 8.5 for 50 years following a fire, assuming that there is no other form of disturbance. We assume that habitat quality follows the pattern in fig. 8.2 and that $r = 2$ (two female offspring per female), $s_a = 0.7$ (70% chance of adult survival), and $s_j = 0.2$ (20% chance of juvenile survival).

and $N(t + 1) = N_a + N_j$. Here,

$$\binom{N}{n} = \frac{N!}{n!(N - n)!}$$

is the binomial coefficient. Since the means of the respective binomial distributions are $s_a N(t)$ and $s_j rq(t)N(t)$, on average and before the next fire, the adult population follows the dynamics

$$N(t + 1) = [s_a + s_j rq(t)]N(t) \tag{8.5}$$

A typical trajectory based on eq. 8.5 makes the trade-off clear (fig. 8.3): following a fire, the population declines but then begins to rise because of the preferred successional habitat. However, as succession proceeds, the habitat becomes less conducive to reproduction, and the population begins to decline. Suppose that we can set fires and that a fraction f of the individuals survive a fire. Depending upon population size, we may be able to increase the longevity of the population by causing a fire and thus resetting the successional process. Thus, we are faced with the problem of having to determine if it is optimal to set a fire.

A dynamic state variable model for optimal fire management

Although extinction is often the criterion used in conservation biology, we take a slightly more conservation-oriented approach and consider a level n_c

above which we'd like to keep the population. It might be zero, in which case we are managing to avoid extinction, or it might be the Allee level. Once the population falls below the Allee level, there is little that can be done (but see the next section).

Thus, we consider two state variables, $N(t) =$ total population size and $Y(t) =$ number of years since the last fire. We set

$$
\begin{aligned}
F(n, y, t) = \text{Maximum probability that the population} \\
\text{stays above } n_c \text{ in every year between } t \text{ and} \\
T, \text{ given that } N(t) = n \text{ and } Y(t) = y
\end{aligned} \tag{8.6}
$$

We continue to use the term *fitness function*, although some other term such as *persistence function* might be more appropriate here. In year T, we have either succeeded or not, so that for every y,

$$
F(n, y, T) = \begin{cases} 1 & \text{if } n > n_c \\ 0 & \text{if } n \leq n_c \end{cases} \tag{8.7}
$$

In addition, in any year we have failed if population size falls to or below the critical level, so that for every y

$$
F(n_c, y, t) = 0 \text{ for } t = 1, 2, \ldots, T \tag{8.8}
$$

In each year, we can either burn or not burn, and we need to compute the fitness value of each decision. Let us begin with the value of not burning, $V_0(n, y, t)$, when $N(t) = n$ and $Y(t) = y$. In that case, $R(t) = rq(y)n$ and $N(t+1)$ follows a distribution given by eq. 8.4. If we know that n_a adults and n_j juveniles survived, then the fitness starting in year $t+1$ would be $F(n_a + n_j, y + 1, t + 1)$. However, we do not know the exact numbers, so we must average over the binomial distributions to obtain

$$
\begin{aligned}
V_0(n, y, t) = \sum_{a=0}^{n} \sum_{j=0}^{rq(y)n} F(a + j, y + 1, t + 1) \binom{n}{a} s_a^a (1 - s_a)^{n-a} \\
\cdot \binom{rq(y)n}{j} s_j^j (1 - s_j)^{rq(y)n-j}
\end{aligned} \tag{8.9}
$$

When a burn is conducted, survival is reduced from s_a or s_j to $s_a f$ or $s_j f$, respectively and the time since the last burn is reset to zero, so that the fitness value of burning is given by

$$
\begin{aligned}
V_b(n, y, t) = \sum_{a=0}^{n} \sum_{j=0}^{rq(y)n} F(a + j, 0, t + 1) \binom{n}{a} (f s_a)^a (1 - f s_a)^{n-a} \\
\cdot \binom{rq(y)n}{j} (f s_j)^j (1 - f s_j)^{rq(y)n-j}
\end{aligned} \tag{8.10}
$$

Then, the dynamic programming equation is

$$F(n, y, t) = \max\{V_0(n, y, t), V_b(n, y, t)\} \tag{8.11}$$

The binomial distributions were computed using the following algorithm.

Computing binomial distributions

Various equations in this chapter require computing binomial distributions of the form

$$\Pr(M, k) = \binom{M}{k} p^k (1 - p)^{M-k}$$

$$= \frac{M!}{k!(M - k)!} p^k (1 - p)^{M-k}$$

Computing these values directly is time-consuming and subject to inaccuracy because the factorial functions can be very large numbers. A much better method is to use an iterative procedure. First note that $\Pr(M, 0) = (1 - p)^M$. Then, note that

$$\Pr(M, k + 1) = \frac{M!}{(k + 1)!(M - k - 1)!} p^{k+1} (1 - p)^{M-k-1}$$

$$= \frac{p(M - k)}{(k + 1)(1 - p)} \Pr(M, k)$$

We use this equation to find $\Pr(M, k + 1)$ for $k = 0, 1, 2, \ldots, M - 1$.

**Envisioning the trade-off between population size
and time since the last fire**

The solution of eq. 8.11 can be used to divide the state plane (y, n) into regions where persistence is maximized by burning and regions in which persistence is maximized by not burning (fig. 8.4). The result is fairly intuitive, once the trade-off is recognized: it is best to burn if the habitat is declining in quality and the population is sufficiently large.

More importantly, the model suggests the kinds of measurements that are needed to make sensible decisions about fire rotations: population size, the time course of habitat quality following a fire, the relationship between habitat quality and reproduction, and the frequency of natural fires. The model also suggests an important operational concept: burns should not be conducted on the basis of fixed rotations but should be conducted in response to population density.

8.3 Managing a metapopulation

Here we discuss how dynamic state variable methods can be used to model the management of a metapopulation.

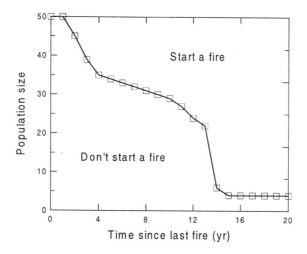

Figure 8.4 The solution of the dynamic state variable model divides the plane into regions where the persistence of the population is maximized by burning and regions where the persistence of the population is maximized by not burning. We show results for the case in which $n_c = 3$, $n_{max} = 50$, $s_a = 0.7$, $s_j = 0.2$, and $f = 0.8$.

A brief history of metapopulation ecology

In the remarkable period of the late 1960s, the theories of optimal foraging (MacArthur and Pianka 1966; Emlen 1966), island biogeography (MacArthur and Wilson 1967), and metapopulation ecology (Levins 1969, 1970) developed. Island biogeography is concerned with the numbers of species occupying a set of islands populated by a source of colonists, whereas metapopulation ecology deals with many populations of one species linked by dispersal. Each subject continues to make contributions to modern ecological science; Rosenzweig (1995) considers them gems of the merging of theory and experiment in biology providing us with a strong predictive ecological theory (of which we do not have many). The theories of optimal foraging and island biogeography developed rapidly and led to experiments and to the development of new fields such as behavioral ecology. On the other hand, metapopulation theory languished for quite a while, and the subsequent development in the 1980s was still mainly theory (e.g., Hanski 1989).

The fundamental variable in Levins's metapopulation model is the fraction p of occupied sites in a homogeneous habitat; the dynamics are assumed to be

$$\frac{dp}{dt} = cp(1 - p) - mp \qquad (8.12)$$

where $cp(1 - p)$ is the colonization rate of patches and mp is the extinction rate of patches. The parameters c and m are rates of local immigration or

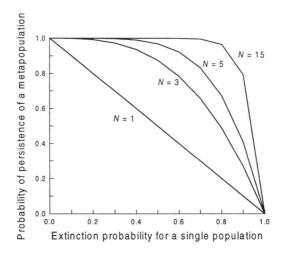

Figure 8.5 The reliability theory notion of a metapopulation is that the collection of populations may persist because of redundancy: all subpopulations have to become extinct simultaneously for the metapopulation to become extinct.

extinction (e.g., Loehle and Li 1996). The steady state level of patch occupancy, which is $1 - m/c$, provided $c > m$, is found by setting the right-hand side of eq. 8.12 equal to zero. Levins's description is based on assumptions of homogeneous patches, no spatial structure, no time lags, and constant probabilities of extinction and immigration (for discussion, see Gotelli 1995).

The Levins model cannot be used to characterize extinction risk of the metapopulation, except in the trivial case in which $m > c$. However, the classic notion of extinction risk in a metapopulation is one based on reliability theory (Martz and Waller 1982). If ρ_e is the probability that one of N independent and identical occupied patches becomes extinct in a year, the probability that all of them do is $(\rho_e)^N$, so that the probability of persistence of at least one patch is $1 - (\rho_e)^N$. This function rises rapidly toward 1 as N increases, as long as the chance of local extinction is not too great (fig. 8.5). Current metapopulation models (e.g., Gilpin and Hanski 1991; Gotelli and Kelly 1993; Adler and Nuernberger 1994; Clark and Rosenzweig 1994; Hanski 1994; Rosenzweig and Clark 1994; Berkson 1996) aim for generality and require tailoring for application. For example, Hanski (1994) introduced a "practical model of metapopulations," and Hanski et al. (1996) and Wahlberg et al. (1996) applied the method to understand the spatial distribution of an endangered butterfly.

Hanski focuses on the **incidence function** J of a patch. This function is the long-term probability that the patch is occupied at any given time, assuming that its dynamics are a first-order Markov process involving colonization and extinction:

$$J_i = \frac{C_i}{C_i + E_i} \tag{8.13}$$

Here, C_i is the colonization probability, E_i is the extinction probability in patch i per unit time, and J_i is the incidence function for patch i.[1]

Hanski assumes that if patch i has area A_i,

$$E_i = \frac{\mu}{A_i^x} \tag{8.14}$$

where μ and x are parameters. Also

$$C_i = \frac{M_i^2}{M_i^2 + y} \tag{8.15}$$

where y is a constant and M_i is the number of immigrants to patch i,

$$M_i = \beta \sum_{j \neq i} p_j \exp(-\alpha d_{ij}) A_j \tag{8.16}$$

where α and β are constants, d_{ij} is the distance from patch i to patch j, $p_j = 1$ if patch j is occupied and zero otherwise, and A_j is the size of patch j. Note that colonization is independent of the number of individuals in other patches and depends only upon the area of the other patches (individuals are presumed to be at constant density) and whether other patches are occupied. The data requirements for this model are stringent (see table 1 in Hanski 1994). Hanski et al. (1996) and Wahlberg et al. (1996) applied this model with reasonable success to butterfly populations. By simulating the metapopulation dynamics, they were also able to study the dynamics of occupancy—that is, the fraction of patches occupied, as a function of time. The fraction of occupied patches depends upon the level of stochasticity in the environment; see fig. 2 of Hanski et al. (1996). Hanski (1994, p. 152) concludes that "Our main task is to make practical yet sensible assumptions about how the extinction and colonization probabilities E_i and C_i depend upon measurable environmental variables and on the life-history traits of the species."

However, this complicated and elegant model provides no advice on improving the persistence of a metapopulation. For example, suppose that one is interested in conserving a butterfly metapopulation. At any particular year, some patches will be empty, and it may be possible to effect recolonization of these patches by transfers from other patches. On the other hand, it may be possible to set up new patches. As with every real problem, resources are limited. Given that only one of the two options (translocation or a new patch) is possible, which is better to do?

---

[1] To derive eq. 8.13, let $J_i(t)$ be the probability that patch i is occupied in year t. Then, $J_i(t+1) = [1 - J_i(t)]C_i + J_i(t)(1 - E_i)$. The incidence function J_i is the value of $J_i(t)$, as $t \to \infty$. Therefore $J_i = (1 - J_i)C_i + J_i(1 - E_i)$. Solving this equation for J_i yields eq. 8.13.

A metapopulation management model

Possingham (1997) analyzes this question with a relatively simple model assuming identical patches. We follow this assumption. There are two state variables in the model

$$N(t) = \text{total number of patches at the start of year } t$$
$$I(t) = \text{number of occupied patches at the start of year } t \qquad (8.17)$$

We describe the state dynamics for $I(t)$ as follows. First, there is a probability μ that an occupied patch will become unoccupied from one year to the next. Second, the probability that an unoccupied patch is colonized depends upon the number of occupied patches. We assume that there is a probability ρ_c that an unoccupied patch is colonized by migration from one of the occupied patches and that colonization from different patches is independent. Thus,

$$\lambda(i) = \text{probability that a given unoccupied patch}$$
$$\text{is colonized in year } t, \text{ given that } I(t) = i \qquad (8.18)$$
$$= 1 - (1 - \rho_c)^i$$

This equation follows from the recognition that $(1-\rho_c)^i$ is the probability that none of the occupied patches colonizes the empty patch under consideration.

Given that $I(t) = i$, the value of $I(t+1)$ can range between zero (no occupied patches) and $N(t)$ (all patches occupied), but the different values of $I(t+1)$ have different probabilities. We find them as follows. First, set

$$a_{ij}(n) = \Pr\{I(t+1) = j | I(t) = i \text{ and } N(t) = n\} \qquad (8.19)$$

We assume that colonization precedes extinction. Suppose that there are k colonizations in year t, when $I(t) = i$. For $I(t+1) = j$, there must be $i+k-j$ extinctions after the colonizations. In addition, k must be constrained so that $i+k-j \geq 0$. (For readers who are uncertain of this logic, it might be helpful to try some numbers: for example, if 8 of 15 patches are colonized this year, what are the different ways of having 10 of the 15 colonized next year?) Because of the assumptions about independence of colonization and extinction, we can use a binomial distribution to characterize the probability of k colonizations followed by $i+k-j$ extinctions:

$$\Pr\{k \text{ colonizations and } i+k-j \text{ extinctions } | I(t) = i, N(t) = n\}$$
$$= \binom{n-i}{k} \lambda(i)^k [1 - \lambda(i)]^{n-i-k} \binom{i+k}{i+k-j} \mu^{i+k-j}(1-\mu)^j \qquad (8.20)$$

We used the following method to compute the binomial coefficients in this expression.

A logarithmic method for computing binomial distributions.

Since $M! = M(M-1)!$ then $\log(M!) = \log(M) + \log((M-1)!)$. Thus, we compute the logarithms of the factorials using $\log(0!) = \log(1!) = 0$ and the iterative procedure to relate $\log(M!)$ to $\log((M-1)!)$. Once these are computed

$$\log[\Pr(M,k)] = \log(M!) - \log(k!) - \log[(M-k)!] + k\log(p) + (M-k)\log(1-p)$$

Since k can take any of the allowed values between zero and $n-i$,

$$a_{ij}(n) = \sum_{k=0}^{n-i} \binom{n-i}{k} \lambda(i)^k [1-\lambda(i)]^{n-i-k} \binom{i+k}{i+k-j} \mu^{i+k-j}(1-\mu)^j \quad (8.21)$$

A check that the $a_{ij}(n)$ have been derived properly is that for every value of i and n,

$$\sum_{j=0}^{n} a_{ij}(n) = 1 \quad (8.22)$$

As an exercise, we encourage you to explain (in words) why this must be true.

A dynamic programming equation for the persistence of the metapopulation

Now we are ready to derive a dynamic programming equation for the persistence of the metapopulation. To do so, we set

$$F(i,n,t) = \text{maximum probability that the meta-} \quad (8.23)$$
$$\text{population persists from } t \text{ to } T, \text{ given}$$
$$\text{that } I(t) = i \text{ and } N(t) = n$$

The maximization here refers to the option of recolonizing some of the uncolonized patches or of establishing new patches. The metapopulation persists at time T if $I(T) > 0$, so that the end condition is

$$F(i,n,T) = \begin{cases} 1 & \text{if } i > 0 \\ 0 & \text{if } i = 0 \end{cases} \quad (8.24)$$

In addition, the boundary condition that the metapopulation is extinct at any time that $I(t) = 0$ is given by

$$F(0,n,t) = 0 \quad (8.25)$$

We assume that in each time period, either a single unoccupied patch can be recolonized or a single new patch created. We also assume that recolonization

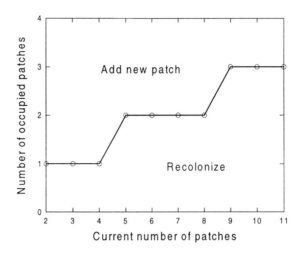

Figure 8.6 The dynamic state variable model generates a boundary for action to maximize the persistence of the metapopulation. We summarize this boundary as a function $I_b(n)$ depending upon the total number of patches. If $i \leq I_b(n)$, the optimal action is to recolonize an empty patch. If $i > I_b(n)$, the optimal action is to add a new patch, unless $n = n_{\max} = 12$, in which case the only option is to recolonize.

of an empty patch succeeds with probability s_c, that a newly created patch gets filled with probability s_f, and that these actions are taken before any of the natural colonization and extinction events. Thus the fitness value of attempting to recolonize a patch when $I(t) = i$ and $N(t) = n$ is

$$
\begin{aligned}
V_c(i, n, t) = s_c \sum_{j=0}^{n} a(i+1, j, n) F(j, n, t+1) \\
+ (1 - s_c) \sum_{j=0}^{n} a(i, j, n) F(j, n, t+1)
\end{aligned}
\tag{8.26}
$$

(Here, we have rewritten $a_{ij}(n)$ as $a(i, j, n)$ to make it easier to read this equation.) The constraint that i cannot exceed n means that the recolonization decision is not available if $i = n$.

The fitness value of creating a new patch is given by

$$
\begin{aligned}
V_p(i, n, t) = s_f \sum_{j=0}^{n+1} a(i+1, j, n+1) F(j, n+1, t+1) \\
+ (1 - s_f) \sum_{j=0}^{n+1} a(i, j, n+1) F(j, n+1, t+1)
\end{aligned}
\tag{8.27}
$$

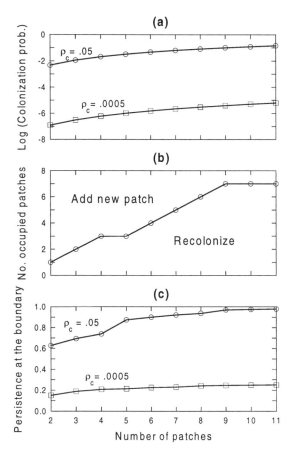

Figure 8.7 (a) Comparison of colonization probabilities for an empty patch when $\rho_c = .05$ (circles) or $\rho_c = .0005$ (squares). These are natural logarithms. (b) The same as in fig. 8.6, but with the reduced value $\rho_c = .0005$. (c) The probability of persistence of the metapopulation over 100 years for $\rho_c = .05$ and $.0005$.

An additional state variable constraint applies here: the number of patches is assumed to be bounded by a maximum $n_{\max}$.

Then the dynamic programming equation is

$$F(i, n, t) = \max\{V_c(i, n, t), V_p(i, n, t)\} \qquad (8.28)$$

The solution of eq. 8.28 separates regions in which the optimal action is to attempt to recolonize an empty patch and others in which the optimal action is to add another patch. For example, we solved it for $n_{\max} = 12$ (maximum of 12 patches), $\rho_c = .05$ (5% chance that an empty patch is colonized by a filled patch), $\mu = .2$ (20% chance that a filled patch becomes empty in one year), s_c = .7 (70% chance that a recolonization succeeds), s_f = .2 (20% chance that a newly made patch fills in the year in which it is made), and $T = 100$ (100 year time horizon). We show the stationary results. In this case, there is a boundary value $I_b(n)$, for the occupied patches (fig. 8.6) for each current number n of patches: if $i \leq I_b(n)$, the optimal action is to recolonize an empty patch. If $i > I_b(n)$, the optimal action is to add a new patch, unless $n = n_{\max}$, in which case the only option is to recolonize.

One can envision using this kind of model to analyze all sorts of trade-offs and sensitivities. For example, the exact amount of dispersal in a metapopulation is often poorly known. When ρ_c = .0005, the chance that an empty patch is colonized by an existing patch is very small (fig 8.7a), and the decisions change considerably (fig 8.7b). Even so, there are plenty of situations in which the optimal decision is to create another patch, rather than to artificially recolonize an existing patch. As one might expect, the long-term persistence of the population is much smaller than before (fig 8.7c). Thus, for example, we might modify the model to allow changes in the colonization probability ρ_c and ask when an increase in it would be the optimal action (but we'll let you do this). Similarly, another action might be to reduce mortality in a particular year, and we might ask when that is the optimal action.

8.4 Some other applications

We close this chapter by noting some other challenges in conservation biology, where dynamic state variable models could be used.

Extinction by hybridization and interspecific mating in western and Clark's grebes

Species of animals and plants may become extinct for many different reasons. As good Darwinians, we recognize that individuals, in the course of breeding, may act in ways that are beneficial for themselves but detrimental to the population. When a rare species is surrounded by an abundant one with which it can mate, the possibility exists for extinction of the rare species via hybridization with the more abundant one. This topic was recently reviewed by Rhymer and Simberloff (1996). Here, we consider a specific example concerned with western and Clark's grebes (Nuechterlein and Buitron 1998).

Western grebes (*Aechmophorus occidentalis*) and Clark's grebes (*A. clarkii*) are monogamous species that maintain pair bonds throughout the season. In both species, there is a slight excess of males in most populations (e.g., 545 M : 455 F at the start of a season). Consequently, as a season progresses and females bond, the opportunities for an unmated male to establish a bond with a member of its own species decreases. Mate selection occurs away from the nest and involves elaborate rituals and mate feeding (Nuechterlein and Storer 1982, 1989); thus it is a property of individual birds, rather than the qualities of their territories. Although Clark's grebe was designated a separate species in 1985, genetic results are somewhat inconsistent (Ahlquist et al. 1987; Eichorst 1994), and mixed pairs of Clark's and western grebes

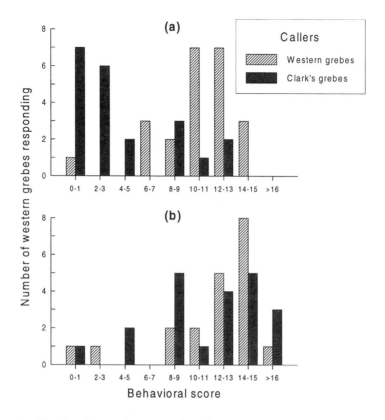

Figure 8.8 Nuechterlein and Buitron (1998) used a playback experiment and behavioral response score to discover that (a) in the middle of the breeding season male western grebes differentially responded to the calls of female western grebes and Clark's grebes ($p < .001$). However, (b) late in the season, the males responded equally to the calls ($p > .10$). These observations led to their formulation of the adaptive hybridization hypothesis.

may be as common as 5% (Nuechterlein and Buitron 1998). The population studied by Nuechterlein and Buitron consisted of more than 3000 individuals, of whom about 95% were western grebes and 5% either Clark's grebes or hybrids. The birds differ in mating plumage and call; individuals use plumage to identify conspecifics (Nuechterlein 1981). However, using playback experiments, Nuechterlein and Buitron showed that as the season progressed, western grebes were more likely to court Clark's grebes (fig. 8.8). Nuechterlein and Buitron (1998) proposed an adaptive hybridization hypothesis: rather than a mistake in species identity, late-season hybrid matings between the two grebe species may be the result of adaptive mate choice by individuals that have limited alternatives. It would be worthwhile to develop a model for this hypothesis and investigate its implications for the conservation of western grebes. Nuechterlein and Buitron suggest that "a bird making a mate choice decision should choose whatever mate provides the most profitable means by which it can pass on its own genes, regardless of whether this leads to greater or less genetic integrity at the level of the gene pool. Thus there should be no selection against hybridization per se, except insofar as it relates to a nonviable means of passing on genes." We lack sufficient biological data to construct a dynamic state variable model for extinction by hybridization, but consider this an ideal opportunity for future work.

Marine reserves

Marine reserves, or marine protected areas (MPAs), are increasingly proposed as a means to protect various stocks of fish and invertebrates and are also proposed as a major rehabilitation tool for depleted stocks (Gubbary 1995; Shackell and Willison 1995; Lauck et al. 1998). However, although the evidence is strong that MPAs achieve benefits within the MPA, the demonstration of benefits outside of the MPAs is weaker.

In such a situation, models can play a valuable role because they allow us to conduct many more investigations than could ever be done empirically. The models can use a suite of transfer functions realistically based in biology, including information on dispersal rates and interception rates, recruitment and the population dynamics of adults, and a careful articulation of objectives. For example, Lauck et al. (1998) used a simple model for population dynamics

$$N(t+1) = N(t)g[N(t)] \tag{8.29}$$

where $g(n)$ is the reproduction per capita. For the discrete logistic, this is

$$g(n) = \exp\left[r(1 - \frac{n}{K})\right] \tag{8.30}$$

where r is the maximum per capita growth rate and K is the carrying capacity.

The concept of an MPA is modeled by assuming that a fraction A ($0 \leq A \leq 1$) of the area in which the stock exists is available for harvesting and that the harvest fraction is targeted at u. Given that the stock at the start of this year is $N(t)$, a fraction $(1-A)N(t)$ remains untouched by the fishery, and a fraction $uAN(t)$ is harvested, so that $(1-u)AN(t)$ escapes. Hence, the remaining stock at the end of fishing is $(1 - A)N(t) + (1 - u)AN(t) = (1 - uA)N(t)$. Lauck et al. assume that the actual harvest fraction fluctuates from its targeted value and follows a beta distribution (see eq. 8.37 below). Note that this simple calculation already indicates an important benefit of the MPA: even if the targeted harvest $\hat{u}$ is greatly exceeded, so that $u \cong 1$, a minimum escapement of $1 - AN(t)$ is assured. Lauck et al. consider the situation in which the reserve size is fixed for the entire time period $t = 1$ to T.

To introduce the idea of protecting the stock, define a critical level n_c, and set

$$F_A(n,t) = \text{probability that the stock stays above the critical level} \tag{8.31}$$
$$n_c \text{ for every year between } t \text{ and } T, \text{ given that } N(t) = n$$
$$\text{and the fraction available for harvesting is } A$$

Then

$$F_A(n,T) = \begin{cases} 1 & \text{if } n > n_c \\ 0 & \text{otherwise} \end{cases} \tag{8.32}$$

For times previous to T, $F_A(n,t)$ satisfies the equation

$$F_A(n,t) = E_u\{F_A(n_u g(n_u), t+1)\} \tag{8.33}$$

where E_u denotes the expectation over the distribution on the harvest fraction and

$$n_u = n(1 - Au) \tag{8.34}$$

The expected total catch can be computed by setting

$$c(n,t) = E_u\left\{\sum_{s=t}^{T-1} u(s)AN(s)\right\} \tag{8.35}$$

so that $c(n,T) = 0$ and

$$\begin{aligned} c(n,t) &= E_u\{Aun + c(n_u g(n_u), t+1)\} \\ &= A\bar{u}N + E_u\{c(n_u g(n_u), t+1)\} \end{aligned} \tag{8.36}$$

where $\bar{u}$ is the mean catch rate. For the distribution on u, Lauck et al. used the beta distribution

$$f(u) = c_n u^{\alpha-1}(1-u)^{\beta-1} \tag{8.37}$$

where α, β are parameters and c_n is a constant such that $\int_0^1 f(u)\,du = 1$. With this density, the mean and the coefficient of variation of u are given by

$$\begin{aligned} \bar{u} &= \frac{\alpha}{\alpha+\beta} \\ CV^2 &= \frac{\beta}{\alpha(\alpha+\beta+1)} \end{aligned} \tag{8.38}$$

Thus, one can specify the mean and coefficient of variation of u and determine the corresponding values of α and β.

Numerical results for this model are presented in Lauck et al. (1998), and in a forthcoming publication of Mangel.

Possingham notes that his examples show "that it is difficult to make generalizations about what is the best management strategy. No simple rules were derived by exploring numerous examples. Every situation needs to be modelled to find the best management strategy ... and searches for generalizations and rules of thumb that provide robust solutions in most situations" (Possingham 1997, p. 397). Thus we anticipate that dynamic state variable models will become additional tools for conservation biologists and population managers.

9

Agroecology

In this chapter, we discuss three applications of dynamic state variable models to problems of agricultural ecology. The first example shows how learning can be incorporated by making information a state variable in a model of parasitoid functional response; we discuss informational questions more fully in chapter 11. The second example shows how dynamic state variable models can be connected to population dynamics of insect parasitoids and their hosts. The third example deals with flowering by thistles, which can be important agricultural pests.

9.1 Behavioral origins of the functional response

As discussed in section 2.8, the functional response is the relationship between an individual predator's rate of consumption and the density of prey. The notion of functional response can be adapted for insect parasitoids, where the interpretation is the relationship between the attack rate of parasitoids and the density of hosts (Holling 1959; Hassell 1978). The classic analysis of functional responses that lead to the Holling disk equation is similar to the rate-maximizing analysis in chapter 4. In this section, we illustrate how more detailed behavioral processes can be incorporated into the construction of the functional response; this is the third method that we identified in section 2.8.

The leaf-miner system

Leaf-mining insects create sinuous linear trails along the leaves of their host plants. For example, the larvae of the moth *Phyllonorycter* make blisterlike mines in the leaves of many trees (Casas 1988). Although the boundaries of the mine are normally visible, the location of the larva within the mine is not. Parasitoids of the larvae (Godfray 1994) find leaves with mines visually during flight (Casas 1989) and use vibrations in the leaf, caused by the movement of the larvae, to find larvae (Vet et al. 1991; Meyhöfer et al. 1994). Although one might envision this as a search game (hiding by the larva and seeking by the parasitoid), we approach it from the viewpoint of learning: when a parasitoid visually detects a mine and lands on a leaf, she does not know the mean time to find the larvae in the mine. As she searches, unsuccessful search provides information about the search rate for this miner on this leaf. Casas et al. (1993) developed a probabilistic model for the functional response of a leaf-miner parasitoid at the behavioral level. Our work complements theirs.

Information and search by the parasitoid

We assume that the parasitoid searches randomly for larvae in mines. Then (Mangel 1987; Hilborn and Mangel 1997),

$$\Pr\{\text{parasitoid detects a larva in a mine after } t \text{ minutes of search}\} = 1 - e^{-bt}$$
$$(9.1)$$

where b is a parameter that characterizes the rate of search; the mean time until the larva is encountered is $1/b$. Equation 9.1 uses the exponential density to characterize the search; this corresponds to random search but leads to the following problem. Assume that the parasitoid has not detected a larva after t minutes of search. To find the probability of not detecting the larva in the next s minutes of search, we apply Bayes' theorem (see appendix to chapter 11)

$$\Pr\{\text{no detection in } t+s \text{ minutes of search} \mid \text{no detection in } t \text{ minutes of search}\}$$

$$= \frac{\Pr\{\text{no detection in } t+s \text{ minutes of search and no detection in } t \text{ minutes of search}\}}{\Pr\{\text{no detection in } t \text{ minutes of search}\}}$$

$$= \frac{\Pr\{\text{no detection in } t+s \text{ minutes of search}\}}{\Pr\{\text{no detection in } t \text{ minutes of search}\}}$$

$$= \frac{e^{-b(t+s)}}{e^{-bt}} = e^{-bs} \qquad (9.2)$$

In other words, the unsuccessful search has no effect on the prediction of the outcome of continued search on this leaf. This is called the memoryless property of the exponential distribution.

A parasitoid that follows this distribution would never leave a particular location because each minute of search is identical! One solution to the problem is to allow the parasitoids to learn during the search. To do this, we assume that there is variation in leaves and mines on leaves. For example, thickness of leaves will affect the transmission of vibrations along the leaf and thus the parasitoid's ability to find the larva. The search success rate is treated as a random variable, which we denote by B, and may take a range of values, which we denote by b. The variation in the search rate before any search is conducted is summarized in the prior frequency distribution, or prior density for $B = b$, which we denote by $f(b)$, with the interpretation,

$$f(b)\Delta b = \Pr\{\text{on a randomly picked leaf } b \leq B \leq b + \Delta b\} \qquad (9.3)$$

For purposes of computation, we use the gamma density

$$f(b) = \frac{\alpha^\nu}{\Gamma(\nu)} e^{-\alpha b} b^{\nu-1} \qquad (9.4)$$

where α and ν are parameters and $\Gamma(\nu)$ is called the gamma function. An explanation of the gamma density for ecologists is given in chapter 3 of Hilborn and Mangel (1997). Briefly, however, since $f(b)$ is a frequency distribution, we require that

$$\int_0^\infty f(b)\, db = 1 \qquad (9.5)$$

that is, the search rate must take some value. Using the right-hand side of eq. 9.4 in eq. 9.5,

$$\int_0^\infty \frac{\alpha^\nu}{\Gamma(\nu)} e^{-\alpha b} b^{\nu-1}\, db = 1 \qquad (9.6)$$

But the first terms in the integral are independent of b and can be taken outside of the integral, so that

$$\frac{\alpha^\nu}{\Gamma(\nu)} \int_0^\infty e^{-\alpha b} b^{\nu-1}\, db = 1 \qquad (9.7)$$

from which we conclude that

$$\int_0^\infty e^{-\alpha b} b^{\nu-1}\, db = \frac{\Gamma(\nu)}{\alpha^\nu} \qquad (9.8)$$

Thus, one can think of $(\Gamma(\nu))/\alpha^\nu$ as the normalization constant for the probability density. Alternatively, eq. 9.8 can be used to define the gamma function:

$$\Gamma(\nu) = \alpha^\nu \int_0^\infty e^{-\alpha b} b^{\nu-1}\, db = \int_0^\infty e^{-x} x^{\nu-1}\, dx \qquad (9.9)$$

(by substituting $\alpha b = x$ in the integral). The most important piece of information that you'll need to know about the gamma function (check Hilborn and Mangel 1997 for further details) is that it satisfies the recursion relationship

$$\Gamma(\nu + 1) = \nu\Gamma(\nu) \qquad (9.10)$$

Thus, if ν is an integer, $\Gamma(\nu+1) = \nu!$ The information in eqs. 9.8–9.10 allows us to find the mean and second moment of the gamma density readily. The mean value of the search rate is $E\{B\} = \nu/\alpha$, and the coefficient of variation of the search rate is $CV\{B\} = 1/\sqrt{\nu}$. It is helpful to imagine that the value of α is determined by the physical attributes of the parasitoid, the leaf, and the mine (smaller α's lead to higher mean search rates) and that the value of ν depends upon the behavior of the larva (smaller values lead to more uncertainty about the search rate on a particular leaf).

With this framework, unsuccessful search provides information about the potential values of the search rate—the probability distribution changes as search continues. This is described in Box 9.1. More important, we are interested in the probability of no detection after t minutes of search, given that there is variation in leaves and mines. If $B = b$, the probability of no detection by time t is e^{-bt}. Since B is unknown, we must average this over the values that it might take

$$\Pr\{\text{no detection by time } t\}$$

$$= \int_0^\infty \Pr\{\text{no detection by time } t | B = b\} f(b)\, db \qquad (9.11)$$

$$= \int_0^\infty e^{-bt} \frac{\alpha^\nu}{\Gamma(\nu)} e^{-\alpha b} b^{\nu-1}\, db$$

Using the hints given, we can simplify the integral and find that

$$\Pr\{\text{no detection by } t\} = \left(\frac{\alpha}{\alpha + t}\right)^\nu \qquad (9.12)$$

and

$$\Pr\{\text{no detection by } t + s | \text{ no detection by } t\} = \left(\frac{\alpha + t}{\alpha + t + s}\right)^\nu \qquad (9.13)$$

Box 9.1 The Parasitoid can learn about search success rates

We apply Bayes' theorem by asking for the posterior density (probability density after the search) for the search rate, given that the search is unsuccessful:

$$f_p(b)\Delta b = \Pr\{\text{on a particular leaf } b \leq B \leq b + \Delta b \mid \text{the search between 0 and } t \text{ was unsuccessful}\}$$

$$= \frac{\Pr\{\text{on a particular leaf } b \leq B \leq B + \Delta b \text{ and the search between 0 and } t \text{ was unsuccessful}\}}{\Pr\{\text{the search between 0 and } t \text{ was unsuccessful}\}}$$

If $B = b$, the probability of unsuccessful search in t minutes is e^{-bt}. Thus, the numerator of this equation is

$$\Pr\{\text{on a particular leaf } b \leq B \leq b + \Delta b \text{ and the search between 0 and } t \text{ was unsuccessful}\} = e^{-bt} f(b)\Delta b$$

The denominator is the chance of unsuccessful search, regardless of the value of B. So, it must be the integral over all values of b of the numerator (alternatively, it is a normalization constant for the numerator). In either case, the denominator is

$$\Pr\{\text{the search between 0 and } t \text{ was unsuccessful}\}$$

$$= \int_0^\infty e^{-b't} f(b') \, db'$$

We use b' in this expression as a different integration variable to avoid confusion.

Combining these equations with eq. 9.4 we find that

$$f_p(b) = \frac{(\alpha + t)^\nu}{\Gamma(\nu)} e^{-(\alpha+t)b} b^{\nu-1}$$

That is, the posterior density is also a gamma density with a changed parameter. In particular now the mean of the search rate is $\nu/(\alpha + t)$—unsuccessful search reduces the estimate of the mean, but does nothing to the level of uncertainty.

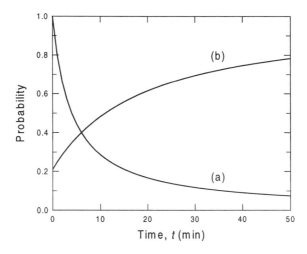

Figure 9.1 When variability in search rate is taken into account, the memoryless property is lost. We show (a) the probability of no detection by t and (b) the probability of no detection by $t + 15$, given no detection by t, when $\nu = 1$ and $\alpha = 4$ min^{-1}. Information about unsuccessful search changes the probability of success in the future.

Comparing eqs. 9.2 and 9.13, we see that the probability of no detection in future search time s is independent of past search time t in eq. 9.2, but not in 9.13. In other words, eq. 9.13 does not have the memoryless property (fig. 9.1).

Incorporating searching into a dynamic state variable model

Now we develop a model for a leaf-miner and its parasitoid. We use two time variables: the time d in the life of the parasitoid and the time t that the parasitoid has already spent looking for a larva on a particular leaf. The question is whether the parasitoid should continue searching on this leaf or move to another one.

To answer this question, we set

$$F(t, d) = \text{maximum expected accumulated fitness for} \qquad (9.14)$$
$$\text{a parasitoid that has already spent time } t$$
$$\text{on a mine at time } d \text{ in the season}$$

Note that t is a state variable and d is the time variable. If the season lasts $D - 1$ time units, then $F(t, D) = 0$ for all values of t. We assume that travel to the next leaf takes τ time units, that oviposition once a larva is found takes 1 time unit, that oviposition provides a fitness increment f_0 to the parasitoid and that the rate of mortality is m. Then the fitness value of moving to a new leaf is given by

$$V_{\text{move}} = e^{-m\tau} F(0, d + \tau) \qquad (9.15)$$

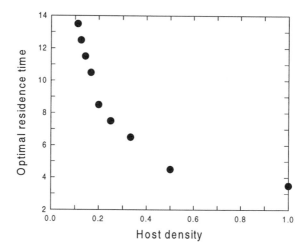

Figure 9.2 Optimal (stationary over values of time within the season) residence times for a parasitoid as a function of host density ($= 1/\tau$). When solving this, we set $D = 300$, $f_0 = 1$, $m = 0$, $\nu = 1$, and $\alpha = 2$.

since the probability of surviving to the new leaf is $e^{-m\tau}$, the parasitoid arrives at the leaf at time $d + \tau$ and has not spent any time searching for the larva on that leaf.

The fitness value of continuing to search is the average, taking into account whether the parasitoid finds the larva during the next period of search. We assume that if she finds the larva, she oviposits (1 time unit) and moves to the next host (τ time units). Thus

$$V_{\text{search}} = \left[1 - \left(\frac{\alpha + t}{\alpha + t + 1}\right)^{\nu}\right]\left[f_0 + e^{-m\tau}F(0, d + \tau + 1)\right]$$

$$+ \left(\frac{\alpha + t}{\alpha + t + 1}\right)^{\nu} F(t + 1, d + 1) \qquad (9.16)$$

The first term on the right-hand side of eq. 9.16 corresponds to the situation in which the parasitoid encounters the larva in the next period of search and the second term in which the next period of search is unsuccessful. The dynamic programming equation is

$$F(t, d) = \max\{V_{\text{move}}, V_{\text{search}}\} \qquad (9.17)$$

The solution of the dynamic programming equation generates the optimal residence time t^* on a leaf, as a function of the parameters and the time d within the season. We focus on the relationship between t^* and τ, the time to find the next leaf. In particular, it is reasonable to assume that τ is inversely related to host density, so for simplicity in presentation, we refer to $1/\tau$ as host density. Solving eq. 9.17 leads to the prediction that optimal residence time decreases as host density increases (fig. 9.2).

Individual behavior and predation rate

A forward iteration allows us to convert from the parasitoid's behavior to the predation rate experienced by the hosts (fig. 9.3). To do this, we fixed the host number at 400 hosts and varied host density by varying τ from 1 to 9. We randomly drew the search rate associated with a host from a gamma density with parameters as in fig. 9.2. If b_j is the search rate associated with the jth host, then the probability that the parasitoid finds it in the optimal search time is $1 - \exp(-b_j t^*)$. By comparing this with a random number, we determine whether the parasitoid successfully finds the jth host. If H is the total number of hosts discovered by the parasitoid, the predation rate in fig. 9.3 is $H/(t^* + \tau)$. Thus we have constructed a functional response from behavioral considerations.

9.2 Behavior and population regulation: The behavioral stabilization of host–parasitoid interactions

The topic of this section—linking behavior and population regulation—has a long history. For example, fig. 1 in Wynne-Edwards's (1962) book shows a high positive correlation between the number of pelagic birds and the abundance of plankton in the North Atlantic Ocean. His interpretation was that the birds respond to the spatial distribution of plankton and that "this is a dispersion that the birds must have brought about by their own efforts" (p. 3)—that is, through their behavior. The data cited by Wynne-Edwards were collected by Jespersen in the 1920s and 1930s; work that broadly connects behavior and fish population dynamics began in the middle of the last century (Smith 1994). In this section, we focus on parasitoids, but clearly the ideas have broad applicability.

Population regulation means bounded fluctuations in abundance or density of populations over time (Murdoch 1994). Hence, we may expect that (1) populations never achieve a steady state, (2) change is common in populations, and (3) populations may be regulated but still have complicated dynamics. Perforce, we must think in terms of probability distributions to characterize the state of populations (Peters et al. 1989). At the same time, the facultative behavior of individuals in their ecological milieu is guided by a genetic program (Mayr 1976) that responds to environmental and internal states and that is subject to natural selection. Thus, behavior is not taken as a given but emerges from the interaction of the environment and the organism: behaviors are the reflections of adaptive responses, and population dynamics are the result of behavior.

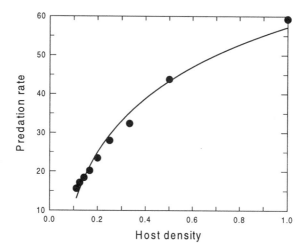

Figure 9.3 The predation rate of an optimally behaving parasitoid, obtained by Monte Carlo forward iteration.

The challenge to the population dynamicist interested in hosts and parasitoids is an old one. The perfectly sensible mathematical model proposed by Nicholson and Bailey (Hassell 1978) leads to biologically nonsensical results, as discussed later. Two solutions to the problem (May 1978; Mangel and Roitberg 1992) will be discussed. We will see that it is precisely because individual behavior occurs at temporal and spatial scales different from those of population dynamics that the study of facultative behavior can enrich population dynamics by providing a means of linking across hierarchical levels in ecological systems (Levin 1992).

Host–parasitoid dynamics between seasons

We consider the dynamics of populations of hosts and parasitoids in which the population numbers of hosts and parasitoids in generation t are $H(t)$ and $P(t)$, respectively. Both are assumed to be univoltine (one generation per year); each host that escapes parasitism produces R offspring in the next year, and each host that is parasitized produces one new parasitoid in the next year. If $f(h, p)$ is the fraction of hosts that escape parasitism when the respective population sizes are h and p, the population dynamics are given by (Hassell 1978)

$$H(t + 1) = R\,H(t)\,f(H(t), P(t))$$
$$P(t + 1) = H(t)\,\{1 - f(H(t), P(t))\}$$

(9.18)

The classical Nicholson–Bailey model is obtained by assuming that $f(h, p) = e^{-ap}$, where a is a parameter. This assumption corresponds to random search, which is equivalent to assuming that the number of hosts attacked follows a Poisson distribution.

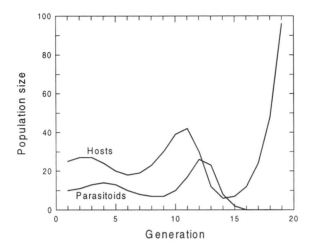

Figure 9.4 The typical result of the Nicholson–Bailey model is increasing oscillations in host and parasitoid population. Depending upon the parameters, the host population may become extinct first (not shown here), in which case the parasitoid population becomes extinct in the next generation. Alternatively (shown here for $R = 2$, $a = 0.06$), the parasitoids may become extinct first, in which case the host population grows exponentially after that.

The problem with this apparently sensible model is that it predicts increasing oscillations (fig. 9.4) with either extinction of the parasitoid and unregulated growth of the host or extinction of both host and parasitoid, depending upon parameters. However, host-parasitoid systems in nature persist with bounded population fluctuations (i.e., regulation) and increasing oscillations are generally not common. During the past sixty years, many workers have tried to resolve this dilemma (Hassell 1978).

May (1978) introduced a phenomenological model by modifying eq. 9.18 according to the logic that the search parameters might vary among parasitoids over space or across time within the season (Chesson and Murdoch 1986; Walde and Murdoch 1988; Murdoch and Stewart-Oaten 1989; Taylor 1991). The particular mechanism of variation is not important here because it occurs within the season and the population dynamics track what happens between the seasons. If the frequency distribution of the attack rate is a gamma density with mean a_0 and coefficient of variation $1/\sqrt{k}$, where k is a parameter called the "overdispersion parameter," then the fraction of hosts that escapes parasitism is found analogously to eq. 9.13 and is given by

$$f(h, p) = \left(\frac{k}{k + a_0 p} \right)^k \tag{9.19}$$

The overdispersion parameter k can be interpreted in terms of the between-patch aggregation of parasitoids. If $k < 1$ (so that the coefficient of variation of attacks is greater than 1), then the standard Nicholson–Bailey model is

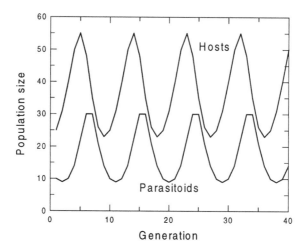

Figure 9.5 When the fraction of hosts that escapes parasitism is given by eq. 9.19 and $k < 1$, the populations no longer have unbounded oscillations. The results here are shown for $R = 2$, $a = 0.06$, and $k = 0.9$. The "flat-top" on the parasitoid population is caused by taking the integer part of the population number that comes from the dynamics in eq. 9.18.

stabilized, so that the populations no longer have unbounded oscillations (fig. 9.5).

The modification in eq. 9.19 was a considerable advance and spawned much research activity (see Walde and Murdoch 1988; Pacala et al. 1990). However, the model provides no mechanism for predicting how k varies according to ecological situations. There is no way to predict, for example, how k will be affected by the distance between patches of hosts, the density of hosts in patches, or the mortality of parasitoids while in patches or traveling between patches.

To answer any of these questions, one must connect the facultative behavior of individual parasitoids with the population dynamics of hosts and parasitoids. We follow the work of Mangel and Roitberg (1992). The key insight is that the equations that characterize the population dynamics, eq. 9.18 or 9.19, relate population sizes at the start of one season to those at the start of the next season, but behavior requires an explicit within-season treatment. The starting point is to reformulate the standard Nicholson–Bailey model as a within-season model by letting $H(s,t)$ and $P(s,t)$ denote the host and parasite populations at time s within season t. The within-season dynamics implicit in the Nicholson–Bailey model are given by

$$\frac{dH(s,t)}{ds} = -\alpha H(s,t)P(s,t)$$

$$\frac{dP(s,t)}{ds} = 0$$

$$(9.20)$$

Integrating these equations from $s = 0$ to $s = S$, the end of the season, shows that the number of parasitoids is steady and thus equal to its value $P(0, t)$ at the start of the season and that the fraction of hosts at the end of the season is $e^{-\alpha S P(0,t)}$. The "constancy" of the parasitoid population does not mean that parasitoids do not die but that births and immigration match death and emigration.

By rewriting the equations to contain within-season dynamics, we can introduce behavior but require two additional kinds of information. First, there must be different behavioral options that provide different fitness payoffs to the parasitoid. Second, one must decide on the suite of variables that characterize the parasitoid's behavior. For the former, we might assume that hosts come in two phenotypes H_1 and H_2, and there is differential parasitoid success in the two phenotypes. There are many sources of such phenotypic variation in the hosts: as larvae, hosts may experience different growth rates and thus be different sizes at the time of attack; or hosts may be in different microenvironments (e.g., sun or shade) and thus, subject to different growth rates and or postparasitism survival rates. The differential fitness payoffs can be characterized by assuming that oviposition in a host of type i produces an offspring with probability p_i, and $p_1 > p_2$.

There are a number of choices for the behavioral variables. In terms of increasing complexity, we might use

- Time within the season: parasitoid oviposition behavior is dynamic and responds to time with the season, so that as the season progresses, the parasitoid shifts from attacking only the better host to attacking either host.
- Time within the season and parasitoid egg complement: the dynamics of oviposition behavior depend upon time within the season and the number of eggs that the parasitoid currently holds in her body (Minkenberg et al. 1992).
- Time within the season, parasitoid egg complement, and experience of encounter rates: in addition to time and egg load, the parasitoid responds facultatively to encounters i.e., she learns (Papaj and Lewis 1993) as she encounters hosts.

The simplest case is the first, in which oviposition behavior is characterized by the time s^* within the season at which the parasitoid starts attacking the less preferred hosts. Assuming that the parasitoid maximizes accumulated reproductive success and that oviposition involves an increased mortality rate, there is a trade-off between current reproduction in a poorer host and expected future reproduction. The time s^* is found by solving a dynamic model for oviposition behavior.

We set

$$F(s) = \text{maximum expected accumulated reproductive} \quad\quad (9.21)$$
$$\text{success of an individual parasitoid from time } s$$
$$\text{to the end of season } S$$

At the end of the season the parasitoid dies, so that $F(S) = 0$. For previous times, we consider the events that can happen to the parasitoid and the behaviors that maximize reproductive success. The events are not encountering any host (no behavioral option), encountering a high quality host (in which case oviposition always produces the highest expected reproductive success), or encountering a lower quality host (in which case oviposition is determined by the balance between current and future reproduction). If λ_1 and λ_2 are the encounter probabilities with the two kinds of hosts, m is mortality while searching and $m_0 > m$ is mortality while ovipositing, then $F(s)$ satisfies the dynamic programming equation

$$F(s) = (1 - \lambda_1 - \lambda_2)(1 - m)F(s + 1) + \lambda_1\{p_1 + (1 - m_0)F(s + 1)\}$$

$$+ \lambda_2 \max\{p_2 + (1 - m_0)F(s + 1); (1 - m)F(s + 1)\} \qquad (9.22)$$

The three terms on the right-hand side of eq. 9.22 correspond to the three encounter events. If $p_2 + (1 - m_0)F(s + 1) < (1 - m)F(s + 1)$, then there is higher fitness payoff from ignoring the host than from ovipositing in it. This means that $s^* > s$. Whenever $s^* > 0$, the facultative behavior of the parasitoids creates a refuge for the less desirable hosts (see Hassell 1978, p. 65).

Finally, we must include the between-season dynamics of hosts and parasitoids and characterize the way that host populations at the start of the next season are related to those at the start of the current season by the fraction of hosts that escape parasitism and by a functional relationship between the total host population and the distribution of the two phenotypes. For example, we might assume that the number of each of the host types is a fixed fraction of the total number of hosts. Alternatively, we might assume that there is a fixed upper limit to the number of better hosts, so that if the total host population is less than this, all hosts are of the better type. Otherwise, the number of higher quality hosts is equal to its maximum, and the remainder of hosts are lower in quality.

Thus, the dynamics are given by

$$\frac{d}{ds}H_1(s,t) = \begin{cases} -\alpha P(0,t)H_1(s,t) & \text{for } s < s^* \\ -\alpha \left[\frac{H_1(0,t)}{H_1(0,t)+H_2(0,t)}\right]H_1(s,t) & \text{for } s \geq s^* \end{cases}$$

$$\qquad (9.23)$$

$$\frac{d}{ds}H_2(s,t) = \begin{cases} 0 & \text{for } s < s^* \\ -\alpha \left[\frac{H_2(0,t)}{H_1(0,t)+H_2(0,t)}\right]H_2(s,t) & \text{for } s \geq s^* \end{cases}$$

When $s < s^*$, hosts of type 1 disappear at a rate proportional to their abundance, and hosts of type 2 do not disappear at all. When $s > s^*$, hosts

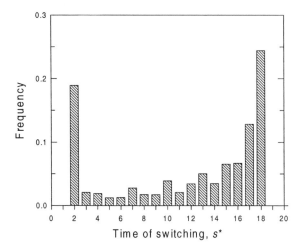

Figure 9.6 An example of the frequency distribution of the time of behavioral switching *s*∗ when the host–parasitoid dynamics were modeled for 10,000 generations (see Mangel and Roitberg 1992, fig. 3 and discussion, for further details).

of each type are parasitized according to their initial abundances. We solve these equations to find that

$$H_1(S,t) = H_1(0,t)\exp[-\alpha P(0,t)s^*]\exp\left[\frac{-\alpha P(0,t)H_1(0,t)(S-s^*)}{H_1(0,t)+H_2(0,t)}\right]$$

$$H_2(S,t) = H_2(0,t)\exp\left[\frac{-\alpha P(0,t)H_2(0,t)(S-s^*)}{H_1(0,t)+H_2(0,t)}\right] \tag{9.24}$$

Note that these equations reduce to the Nicholson–Bailey model under appropriate simplifying assumptions (e.g., host type 2 is never attacked, or host types are identical).

The total number of unparasitized hosts at the end of the season is given by $H_T(S,t) = H_1(S,t) + H_2(S,t)$. Assuming that they contribute equally to the production of offspring, we find that the total number of hosts at the start of the next season is

$$H_T(0,t+1) = R[H_1(S,t) + H_2(S,t)] \tag{9.25}$$

Similarly, the parasitoid population at the start of the next season is composed of parasitoids that emerge from host type 1 and those that emerge from host type 2. Assuming that the latter are less likely to emerge than the former by a factor of β,

$$P(0,t+1) = [H_1(0,t) - H_1(S,t)] + \beta[H_2(0,t) - H_2(S,t)] \tag{9.26}$$

The description of host population dynamics will be complete once we specify how the numbers of host types 1 and 2 are related to the total host population at the start of the season. Mangel and Roitberg (1992) assumed that there is a potentially nonlinear relationship between $H_i(0, t+1)$ and $H_T(0, t+1)$. The simplest case is density-independent production of phenotypes, in which the different kinds of hosts are produced in constant proportions. The alternative is density-dependent production of host phenotypes. A specific model of density-dependent production of phenotypes is given by Mangel and Roitberg (1992).

As shown by Mangel and Roitberg (1992), when facultative within-season behavior of parasitoids occurs as described here, the host and parasitoid populations are stabilized. Oscillations in population numbers persist, but these oscillations remain bounded and neither population becomes extinct. Certain parameter combinations lead to limit-cycle oscillations, whereas other parameters result in other kinds of attractors.

Three main features emerge from this work. First, within-season individual behavior can stabilize host–parasitoid population dynamics. Second, depending upon the way parasitoids respond to host encounter rates, the dynamic behavior can be quite complex (Mangel and Roitberg 1992). Third, there may be considerable variation in s^* between seasons (fig. 9.6). Each value of s^* in fig. 9.6 is exactly predictable from the initial density of hosts and parasitoids at the start of the season, using the dynamic model that describes parasitoid behavior within the season, and from the rule that relates the host phenotypes to the total host population. The variation in s^* is caused by the changes in behavior that are linked to changes in host and parasitoid populations. Were one to simply observe this variation without understanding its origins, there might be a tendency to attribute it to "noise" rather than biology. To do this would clearly be wrong.

9.3 Weed control: Predicting flowering in thistles and other monocarpic perennials

Many thistles such as the beautiful bull thistle *Cirsium vulgare* are important agricultural pests (see Louda and Potvin 1995; Rees et al. 1999 for reviews). Many of them are also monocarpic perennials: they live for more than one year, flower once, and die. For such plants, it is important to be able to predict the age or size at flowering (see Klinkhamer et al. 1987; de Jong et al. 1989; Kachi 1990; Bullock et al. 1994), because knowing this affects plans to control the weeds.

A generic life history for a thistle goes as follows: after the seed germinates and the seedling establishes itself, a rosette is formed. Typically, these are low and flat leaf structures. In the year that a thistle flowers, it bolts in the spring, sending up a stalk, and then produces flowers throughout the summer. Then the problem is simple: given rosette size and age this year, can we predict whether the plant will flower or not next year? Rees et al. (1999) answered this question; we follow their analysis with some change in notation.

Dynamics of growth, reproduction, and survival in monocarpic perennials

We characterize the state of the plant by some measure of size (e.g., rosette diameter, leaf area) at the time reproduction is initiated. If $L(t)$ denotes the logarithm of this size at the start of growing season t, many empirical papers (e.g., Kachi and Hirose 1985; de Jong et al. 1989; Rees et al. 1999) suggest that growth of the logarithm of size is linear so that

$$L(t+1) = a_g + b_g L(t) + Z(t) \tag{9.27}$$

where a_g and b_g are growth parameters and $Z(t)$ represents a random microsite (soil variation, herbivory) component of growth. We treat it as the residual in regression analysis of plots of $L(t+1)$ versus $L(t)$.

For the plant to grow, we require that $a_g > 0$, and for it to achieve an asymptotic size, that $b_g < 1$. Then the asymptotic size L_∞ on a log scale is found by setting

$$L_\infty = a_g + b_g L_\infty \tag{9.28}$$

so that

$$L_\infty = \frac{a_g}{1 - b_g} \tag{9.29}$$

We assume that $Z(t)$ is normally distributed with mean zero and standard deviation $\sigma(L)$ that depends linearly upon length

$$\sigma(L) = a_\sigma + b_\sigma L \tag{9.30}$$

Various data suggest that $b_\sigma < 0$, so that larger plants experience less variation in growth rates. To insure that the standard deviation remains positive, we replace eq. 9.30 by

$$\sigma(L) = \max(a_\sigma + b_\sigma L, \sigma_{\min}) \tag{9.31}$$

where $\sigma_{\min}$ is estimated by the minimum observed standard deviation in growth rate at maximum length.

When a plant of size L reproduces, seed set $R(L)$ is often a power function of size, which we can represent as an exponential

$$R(L) = \exp(A + BL) \tag{9.32}$$

where A and B are allometric parameters relating log (seed set) to log (size) (van der Meijden and van der Waals-Kooi 1979; Reinartz 1984; Kachi and Hirose 1985; Samson and Werk 1986; Rees and Crawley 1989; Schat et al.

1989; Klinkhamer et al. 1992; Rees et al. 1999). We understand that $R(L) < 1$ means that no seeds are produced.

Survival in many plants is strongly size-dependent, but the age-dependent component, although statistically significant, is a relatively poor predictor of plant fate (Werner 1975; van der Meijden and van der Waals-Kooi 1979; Gross 1981; Rees et al. 1999). Thus, we characterize the survival of a plant of size $L(t)$ to the start of year $t+1$ only by its size; growth takes place before mortality (Rees et al. 1999). The survival function we use is

$$\Pr\{\text{survive from } t \text{ to } t+1 | L(t)\} = 1 - \exp[-s_0 - s_1 L(t)] \qquad (9.33)$$

with size-independent parameter s_0 and size-dependent parameter s_1.

Determination of the size and age at flowering

We want to predict the size and age at which individuals will flower, given the growth dynamics, size-dependent survival, and size-dependent reproduction. To do this, we set

$$F(\ell, t) = \text{the maximum expected seed production} \qquad (9.34)$$
$$\text{of a plant of size } L(t) = \ell \text{ at age } t$$

We assume that there is an age T at which the plant must reproduce, so that

$$F(\ell, T) = \exp(A + B\ell) \qquad (9.35)$$

The terminal age T can be interpreted alternatively as the time of reproductive senescence or the time at which successional changes make reproduction mandatory.

For ages previous to T, $F(\ell, t)$ is determined by comparing the value of current reproduction with the expected value of future reproduction, taking growth and survival into account. The fitness value (measured in terms of expected reproduction) $V_{\text{now}}(\ell, t)$ of reproducing at age t for a plant of size ℓ is given by

$$V_{\text{now}}(\ell, t) = \exp(A + B\ell) \qquad (9.36)$$

Assuming that growth takes place before risk of mortality, the fitness value of continuing to grow one more year is given by

$$V_{\text{grow}}(\ell, t) = E_Z \left[1 - \exp(-s_0 - s_1(a_g + b_g\ell + Z)) \right] F(a_g + b_g\ell + Z, t+1) \qquad (9.37)$$

and the dynamic programming equation is

$$F(\ell, t) = \max\{V_{\text{now}}(\ell, t), V_{\text{grow}}(\ell, t)\} \qquad (9.38)$$

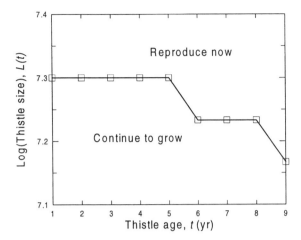

Figure 9.7 The dynamic state variable model predicts a switching curve for reproduction in thistles. Plants continue to grow if their size at a specific age falls below the switching curve; otherwise they reproduce. Minimum standard deviation = 0.5.

We illustrate the solution using data for one of the sites in France (Crau) studied by Rees et al. (1999). For this site, the parameters are $T = 10$, $A = -2.36$, $B = 1.29$, $a_g = 3.234$, $b_g = 0.56$, $s_0 = 0.11$, $s_1 = 0.16$, $a_\sigma = 1.57$, and $b_\sigma = -0.16$. Thus the asymptotic size is $L_\infty = 7.35$.

The solution of eq. 9.38 generates a switching curve $L_s(t)$. If the plant's size is above this curve, we predict that it will reproduce; otherwise it is predicted to continue growing (fig. 9.7). Note that as thistles age, the switching value for reproduction decreases. This can be understood as follows. The deterministic component of growth is limited because $b_g < 1$ and growth will slow down as size gets closer to asymptotic size. However, a sequence of "lucky" years with the "right" random components of growth may allow a plant to exceed the asymptotic size in one year, and thus it will be reduced in the next year. As the plant approaches T, the opportunity for such lucky growth declines, and thus the switching curve declines. Recall too, that $L(t)$ is the logarithm of rosette area. Thus, the asymptotic size (7.35), maximum switching value (7.30), and minimum switching value (7.167) correspond to 1556 cm^2, 1480 cm^2, and 1296 cm^2, respectively.

Consonant with the notion that switching is intimately connected with the stochastic component of growth, we find that the switching value increases as the minimum standard deviation increases (fig. 9.8). The observation of Rees et al. (1999) was that the standard deviation of the stochastic component of growth decreased with increasing plant size, and we captured that effect with eq. 9.31. Larger values of σ_{min} will provide greater growth increments with higher probability. Hence, we expect that the switching curve will increase as σ_{min} increases.

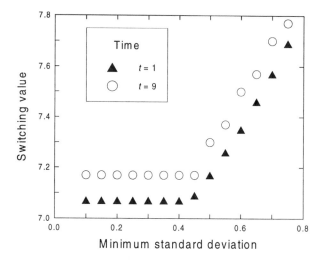

Figure 9.8 The switching value increases as the minimum standard deviation increases. We show the switching value at $t = 1$ (triangles) and $t = 9$ (circles).

9.4 Discussion

We briefly discuss two other areas of agroecology where dynamic state variable models can contribute.

The first involves linking facultative behavior and animal movement. For example, Morris (1993) studied pollen dispersal by honeybees moving on a linear array of flowers. He measured the residence time τ of bees on flowers, the distance δ_L or δ_R that they moved to the left or right when they made a move, and the probability p that they moved to the right. These parameters can be combined as the advection velocity c and the diffusion coefficient D in a diffusion model (Okubo 1980; Murray 1990) that characterizes the probability that at time t the bee will be located at a specified spatial point:

$$c = \frac{p\delta_R - (1 - p)\delta_L}{\tau}$$

$$D = \frac{p(\delta_R)^2 + (1 - p)(\delta_L)^2}{2\tau} \tag{9.39}$$

Diffusion models are commonly used in both basic and applied studies (Corbett and Plant 1993; Taylor 1991; Turchin and Thoeny 1993).

Morris noted that c and D are connected to individual behavior, so that by measuring the behaviors of individual bees—the duration of visits and the probability of moving in each direction—as a function of plant spacing, he was able to estimate the parameters. Then Morris used the resulting diffusion model to predict the spread of pollen from the initial point. In subsequent work (Morris et al. 1995), the movement model was combined with models for pollen deposition to evaluate the actual pollen deposition, as a function of space and time. This work has important implications for the spread of engineered genes by pollinators.

However, the limitation of the diffusion model is that it provides no method for predicting the parameters τ, δ_R, δ_L, and p or the derived parameters c and D without conducting another experiment. A behavioral model would allow one to make such predictions, and this would have many important implications, especially if we want to understand how pests and natural enemies will respond to diversification in agroecosystems (Corbett and Plant 1993; Messing et al. 1994; Turchin and Thoeny 1993).

The second example involves habitat choice by flour beetles *Tribolium* spp. (Sokoloff 1974; Costantino and Desharnais 1991), which are important pests of stored grains. As the beetles use flour, they "condition it," and the flour turns grayish. Some authors believe that quinones, which are beetle waste products, might be the key conditioner of the flour. Thomas Park observed (Sokoloff 1974, p. 150) that conditioned flour led to reduced cannibalism, lowered fecundity, increased duration of the larval period, and increased mortality of eggs, larvae, and pupae.

Beetles behaviorally respond to conditioned flour. In general, the oviposition rate decreases as the conditioning of the flour increases. When given an option, beetles do not enter conditioned flour and will disperse from media heavily saturated with quinones. Although dispersal studies of beetles began with Park and his colleagues, the work has generally been at the population level, with little focus connecting individual behavior and population dynamics. However, the system is just about perfect for behavioral studies. Flour beetles spend their entire lives in flour, and it is possible to accurately characterize the demographic parameters (birth and death rates), as a function of the numbers of individuals and the quality of the flour (Sokoloff 1974; Costantino and Desharnais 1991). Thus, one can connect reproductive success of individuals and dispersal by first, titrating reproductive success as a function of quinone in flour, second, constructing a model of individual behavior (e.g., stay in the current patch or seek another one), and third, using forward iteration to predict population consequences. Then these detailed predictions can be tested in the laboratory (which is the "natural" setting for *Tribolium*) in a variety of different media and in a variety of different physical settings.

10

Population-Level Models

> **Techniques and concepts used in chapter 10**
>
> Evolutionarily stable strategies (10.1, 10.2)
> Forward iteration using Markov chains (10.2)
> Repeated backward and forward iterations (10.2)
> Errors in decisions (10.2)
> Genetic algorithms (10.3)

Most of the dynamic state variable models described in this book pertain to individual optimization: What is optimal for one individual is assumed to be independent of the way other individuals are behaving. In this chapter we indicate briefly how dynamic state variable modeling can be extended to deal with population-level phenomena by using the example of state-dependent ideal free distributions. We also discuss an alternative approach to dynamic optimization based on the method of genetic algorithms.

10.1 Ideal free distributions

Individual optimization models ignore the possibility that behavior of other members in a population of similar organisms may affect the fitness of a given individual. Let us reconsider, for example, the patch-selection model of chapter 1; here we assume that all three patches are used for foraging. In this model it is assumed that the forager chooses between patches characterized by fixed predation risks and foraging benefits. By using forward iteration, we can calculate the average proportion of time that the forager spends on each patch. For a population of N foragers, this translates into the average number of foragers that use each patch at any time. Suppose, for example, that 10% use patch 1, 20% use patch 2, and 70% use patch 3. In practice this concentration of foragers could have two effects. The intake rate in patch 3 could become reduced, either because of depletion of resources, or because of frequent competitive interactions, or both. Predation risks on patch 3 could increase because predators tend to concentrate where their prey are more abundant, or could decrease because of dilution effects (Hamilton 1964). Both effects would change the parameters of the model, which would in turn alter the optimal patch choices, and lead to a redistribution of the N foragers among the three patches. This is a case of **frequency dependence**, in the sense that the optimal decision for an individual depends on what other members of the population are doing. Wallace (1982) and Schlichting and Pigliucci (1998, p. 336) argue that frequency-dependent selection, rather than density-dependent selection, is the more common selective regime.

One of the earliest models of frequency-dependent foraging is the ideal free distribution model of Fretwell and Lucas (1970). This model assumes that a population of foragers will distribute themselves among k patches to equate net foraging benefits from each patch. Implications of the ideal free distribution for population ecology are discussed by Lomnicki (1988) and Sutherland (1996). Now the ideal free distribution is recognized as an example of an **evolutionarily stable strategy** (ESS) in the sense of Maynard Smith (1982). An ESS is defined as a behavioral strategy S that has the property that when all members of a population employ S, the optimal strategy for any new, or mutant individual is also S. In other words, the strategy S cannot be successfully invaded by any alternative strategy S'. If the ideal free distribution is interpreted as specifying the frequencies with which an individual uses each patch, then the ideal free distribution is an ESS in this sense. State-dependent ideal free distributions are discussed by McNamara and Houston (1990).

Important extensions of the ideal free distribution model that include predation risk, as well as foraging benefits, were discussed by Schwinning and Rosenzweig (1990) and by Hugie and Dill (1994). These authors considered the simultaneous ideal free distributions of mobile predators and their prey. As pointed out by Schwinning and Rosenzweig (see also Mangel 1990a), the resulting ideal free distribution is usually dynamically unstable, and goes through periodic oscillations: if prey concentrate on a given patch, the predators will tend to concentrate there, too. This makes the patch undesirable for the prey, which then move to a different patch, only to be followed by the predators, and so on. Some habitat-choice experiments suggestive of oscillatory dynamics are cited in the paper of Schwinning and Rosenzweig, who also discuss three ways in which the oscillations could be stabilized: (1) protection provided by a refuge patch, (2) interspecific competition among prey, and (3) a threshold of fitness difference between patches, below which the prey do not change patches.

In the next section, we will use the patch-selection model to bring these various ideas together. We will also demonstrate the technique of repeated backward and forward iteration as a method of computing evolutionarily stable strategies in frequency-dependent dynamic state variable models.

Table 10.1 Parameter values for the patch-selection model

Patch (i)	Predation risk (m_{i0})	Probability of food (λ_i)	Size of food item (Y_i)
1	0	0	0
2	0.004	0.4	3
3	0.02	0.6	5

10.2 Patch choice with frequency dependent predation risk

Consider again the patch-selection model of chapter 1, ignoring reproduction for simplicity. We assume the parameter values in table 10.1.

Metabolic cost per period is $c = 1$ unit and the state variable constraints are $0 \leq X(t) \leq x_{\max} = 10$. Now we will add the additional assumption that the predation risk in patches 2 and 3 depends on the fraction f_i of the total population N of foragers that use each patch. Thus, we imagine that predators tend to concentrate where prey are most abundant, so that

$$\text{Predation risk on patch } i = m_{i0} f_i \qquad (10.1)$$

where

$$f_i = \frac{N_i}{N_2 + N_3} \quad (i = 2, 3) \qquad (10.2)$$

with N_i = number of foragers located in patch i. Notice that $f_2 + f_3 = 1$.

If f_2 is given (so that $f_3 = 1 - f_2$), we can find the optimal decision matrix $D(x, t, f_2)$, depending on f_2, by solving the dynamic programming equation (backward iteration) as before:

$$F(x, t) = \max\{V_1(x, t), V_2(x, t), V_3(x, t)\} \qquad (10.3)$$

where

$$V_i(x, t) = (1 - m_{i0} f_i)[\lambda_i F(x + Y_i - c, t + 1) + (1 - \lambda_i) F(x - c, t + 1)] \qquad (10.4)$$

with terminal condition

$$F(x, T) = \begin{cases} 1 & \text{if } x > 0 \\ 0 & \text{if } x = 0 \end{cases} \qquad (10.5)$$

We will consider only the stationary strategy $D_s(x; f_2)$, equal to $D(x, t; f_2)$ when $t \ll T$. Next we use forward iteration to compute the stationary distribution of the state variable $X(t)$, for a population of foragers that uses the strategy $D_s(x; f_2)$. Rather than Monte Carlo forward iteration (an approximate method), we use the method of **Markov chains** (an exact method), which allows us to compute the distribution of the state variable X, previously obtained by forward simulation. (See Karlin and Taylor 1975 for an introduction to Markov chains.) First, we find the Markov transition matrix $M = (M_{xy})$ where

$$M_{xy} = \Pr\{X(t+1) = y | X(t) = x\} \tag{10.6}$$

To calculate these transition probabilities, we proceed as follows:

1. Since dead organisms remain dead, we set $M_{00} = 1$ and $M_{0y} = 0$ for $y = 1, 2, \ldots, x_{max}$.
2. For each $x = 1, 2, \ldots, x_{max}$, we set $i = D_s(x; f_2)$. Since we can ignore the case in which the forager is killed by a predator (because we only need to compute the proportions of surviving foragers that use each patch, as explained below),

$$M_{xy} = \begin{cases} \lambda_i & \text{for } y = x + Y_i - c \\ (1 - \lambda_i) & \text{for } y = x - c \\ 0 & \text{for all other values of } y \end{cases} \tag{10.7}$$

with the constraint that $0 \leq y \leq x_{max}$.

Now consider a population of individuals that uses the stationary strategy $D_s(x; f_2)$. Let

$$\pi_x(t) = \Pr\{\text{individual has } X(t) = x\} \tag{10.8}$$

Then,

$$\pi_x(t+1) = \Pr\{\text{individual has } X(t+1) = x\}$$
$$= \Sigma_z \Pr\{\text{individual has } X(t+1) = x | X(t) = z\} \cdot \Pr\{X(t) = z\}$$
$$= \Sigma_z M_{zx} \pi_z(t) \tag{10.9}$$

This is the **forward iteration equation** of the Markov chain. Suppose that all individuals start with $X(1) = x_1$; then

$$\pi_x(1) = \begin{cases} 1 & x = x_1 \\ 0 & \text{all other } x \end{cases}$$

Then eq. (10.9) is iterated to obtain $\pi_x(t)$ for $t = 2, 3, \ldots$. By the theory of Markov chains, it happens that $\pi_x(t)$ converges to a stationary distribution $\overline{\pi}_x = \lim_{t \to \infty} \pi_x(t)$. We ran the iterations of eq. 10.9 for 50 time periods, displaying the computed values for $t = 49$ and 50 to check that indeed

$\pi_x(49) \cong \pi_x(50)$—that is, the stationary distribution has been reached.

Thus, after sufficiently many time periods, the probability that an individual has state $X(t) = x$ is equal to $\overline{\pi}_x$. From this, we can calculate the distribution of foragers (from a population of size N) located on each of the three patches. To do this, we set

$$u_i(x) = \Pr\{\text{individual uses patch } i | X(t) = x\} :$$
$$= \begin{cases} 1 & \text{if } i = D_s(x; f_2) \\ 0 & \text{otherwise} \end{cases} \tag{10.10}$$

Then,

$$
\begin{aligned}
N_i &= \text{number using patch } i \\
&= N \cdot \Pr\{\text{individual uses patch } i\} \\
&= N \cdot \Sigma_x \Pr\{\text{individual uses patch } i | X(t) = x\} \Pr\{X(t) = x\} \\
&= N \cdot \Sigma_x u_i(x) \overline{\pi}_x
\end{aligned}
\tag{10.11}
$$

Finally, the proportion of the population that uses patch 2, out of those using either patch 2 or patch 3, is given by

$$f_2' = \frac{N_2}{N_2 + N_3} \tag{10.12}$$

(The total population size N cancels out in this ratio and therefore can be ignored in the calculation.)

For example, suppose we assume that $f_2 = 0.5$—that is, the foragers are distributed equally over patches 2 and 3 (some foragers may be using patch 1, but this doesn't affect the results). Carrying out the calculations described, we obtain

$$
\begin{aligned}
D_s(x; f_2) &= (3, 3, 3, 3, 3, 3, 3, 2, 2, 1) \\
\overline{\pi}_x &= (0, 0, 0, 0, .01, .02, .04, .10, .24, .59) \\
u_2(x) &= (0, 0, 0, 0, 0, 0, 0, 1, 1, 0) \\
u_3(x) &= (1, 1, 1, 1, 1, 1, 1, 1, 0, 0, 0)
\end{aligned}
$$

for $x = 1, 2, 3, \ldots, 10$. Thus the distribution of foragers over the two patches is $N_2 = 0.34N$ and $N_3 = 0.07N$, and therefore $f_2' = N_2/(N_2 + N_3) = 0.829$.

Finding the ESS

There is clearly an inconsistency in this result. If we start by assuming that $f_2 = 0.5$—that is, foragers are equally distributed between patches 2 and 3, we conclude that the optimal patch choice implies instead that foragers are

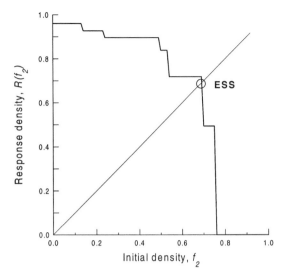

Figure 10.1 The response curve $f_2' = R(f_2)$ for the patch-selection model with frequency-dependent predation risk. The ESS $\overline{f}_2$ is determined by $\overline{f}_2 = R(\overline{f}_2)$.

distributed 82.9% on patch 2 and 17.1% on patch 3. We say that $f_2' = 0.829$ is the optimal **population-level response** to the assumed value $f_2 = 0.5$. For consistency, we should clearly require that $f_2' = f_2$. If this were the case, the population would have no tendency to redistribute itself between patches 2 and 3. In other words, this consistent value f_2 (if it exists) defines the ideal free distribution, or the ESS, for the model. That is, once we have calculated the consistent value of f_2, then we can calculate the optimal decisions $D_2(x; f_2)$. If the entire population uses these decisions, then the optimal decision for any individual is also given by $D_2(x; f_2)$. (In this model, the calculation of the ESS reduces to finding a single consistent density f_2. In more complicated models, this may not be the case; we discuss this possibility further later.) How can we calculate the consistent (ESS) value of f_2? Let us write

$$f_2' = R(f_2) \tag{10.13}$$

(R for "response to"). We can use our code to calculate $R(f_2)$ for values of f_2 that range from zero to one—see fig. 10.1. The consistent value $\overline{f}_2$ satisfies $R(\overline{f}_2) = \overline{f}_2$; as seen from the figure,

$$\overline{f}_2 = 0.69$$

In this example, we have succeeded in calculating the ESS graphically, using eq. 10.13. Our success depended on having a single parameter f_2 that specifies the population's distribution among two foraging patches. More complicated models might involve several such parameters. For example, a patch-selection model that has three foraging patches would have two independent parameters f_2, f_3 (with $f_4 = 1 - f_2 - f_3$), and the best response equation would be

$$f_2' = R_2(f_2, f_3)$$
$$f_3' = R_3(f_2, f_3)$$
(10.14)

These values can be calculated by the process of backward and forward iteration described before. The ESS would be given by

$$\bar{f}_2 = R_2(\bar{f}_2, \bar{f}_3)$$
$$\bar{f}_3 = R_3(\bar{f}_2, \bar{f}_3)$$
(10.15)

but it is not immediately obvious how to solve this system of simultaneous equations. One method, which is elegant and easy to use—if it works—employs iteration, namely, we start with initial values f_2, f_3 and calculate the best response f_2', f_3'. Next, we repeat (iterate) this procedure and calculate f_2'', f_3'', and so on. If we are lucky, these iterations will converge to values $\bar{f}_2, \bar{f}_3$, which will necessarily satisfy the ESS condition, eq. 10.15.

There is no guarantee, however that these iterations will actually converge. Indeed, for the patch-selection model they do not converge, even for the original model with only one parameter f_2. Figure 10.2(a) shows the successive iterates $f_2' = R(f_2)$, $f_2'' = R(f_2')$, etc., for this model; here, the response curve is as in fig. 10.1. The actual values are $f_2' = 0.9$, $f_2'' = 0.0$, $f_2''' = 0.95$, $f_2^{iv} = 0.0$, and so on. Clearly these iterations fail to converge.

The biological explanation for this nonconvergence is related to the dynamic oscillations in the ideal free distribution of mobile predators and prey, as noted by Schwinning and Rosenzweig (1990). When $f_2 < \bar{f}_2$ few foragers use patch 2, and predation risk is low there. Thus, foragers will prefer patch 2, so that $f_2' > f_2$. In fact, $f_2' > \bar{f}_2$ (see fig. 10.2) because of the fact that the response curve $R(f_2)$ has a negative slope. This necessarily leads to oscillations of the iterated values $f_2', f_2'', \ldots$ about $\bar{f}_2$. However, it is not necessary that these oscillating values fail to converge; fig. 10.2(b) shows a situation in which they converge instead to the ESS $\bar{f}_2$. What determines whether we get convergence or not is the slope of the response curve at $\bar{f}_2$: if this slope is steeper than 45° (i.e., $R_2'(\bar{f}_2) < -1$) the oscillations fail to converge, and vice versa. This is illustrated in fig. 10.2, panels (a) and (b).

Errors in decision making

McNamara et al. (1997) have shown how to eliminate oscillations of this kind in ESS calculations by assuming **errors in decision making**. First, to consider an extreme case, suppose in the patch-selection model that the forager chooses patches at random, rather than according to the optimal strategy. Then, the value of f_2' is independent of f_2, and the forager's behavior does not take

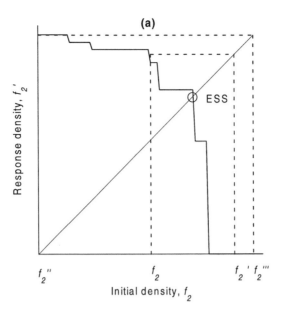

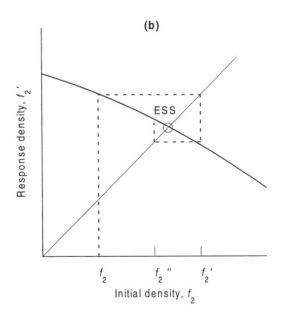

Figure 10.2 Cobweb diagrams for calculating the ESS $\overline{f}_2$ by iteration: $f_2' = R(f_2)$, $f_2'' = R(f_2')$, etc. (a) The iteration method fails, that is, these iterations fail to converge; (b) The iterations converge to the ESS $\overline{f}_2$.

account of predation risks. In this case, the response curve is a horizontal line, and the iterations converge immediately to the ESS (as you can see by imagining a level response curve in fig. 10.2b). Less extreme errors lead to a response curve that is not flat but less steep than the curve in fig. 10.2(a).

To be more explicit about how errors in decision making can be used to model population responses, note that if $V^*(x, t)$ is the fitness value of choosing the optimal patch when $X(t) = x$ and $V_i(x, t)$ is the fitness value of choosing patch i, then,

$$C_i(x, t) = V^*(x, t) - V_i(x, t) \qquad (10.16)$$

is the cost of choosing patch i. McNamara and Houston (1986) called this the canonical cost; it is the loss in fitness that results from an erroneous decision.

Now we introduce a function $h(c)$ that has the properties that

$$h(0) = 1 \text{ and } h(c) \text{ decreases as } c \text{ increases} \qquad (10.17)$$

and we assume that

$$\Pr\{\text{forager uses patch } i\} = \frac{h(C_i)}{\sum_j h(C_j)} \qquad (10.18)$$

Thus the more costly an error is, the less likely it is to be made by the forager. For the computations, we take

$$h(c) = \frac{c_0}{c + c_0} \qquad (10.19)$$

where c_0 is a constant that has the effect that $h(c_0) = 0.5$. Thus, when c_0 is small, even small errors tend to be avoided, whereas when c_0 is large, errors are more prevalent. The values of c_0 used in the calculations determine the numerical stability of the iterations, as we demonstrate now.

To carry out the calculations for our patch-selection model, first we obtain the backward iteration algorithm based on the modified fitness function (see section 2.6).

$$\tilde{F}(x, t) = \text{probability the individual survives from} \qquad (10.20)$$
$$t \text{ to } T, \text{ given that } X(t) = x, \text{ and the}$$
$$\text{individual makes decisions with errors}$$
$$\text{according to condition 10.18}$$

Then,

$$\tilde{F}(x, T) = \begin{cases} 1 & \text{if } x > 0 \\ 0 & \text{if } x = 0 \end{cases} \qquad (10.21)$$

and for $t < T$

$$V_i(x,t) = (1 - m_{i0}f_i)[\lambda_i \tilde{F}(x + Y_i - c, t + 1) + (1 - \lambda_i)\tilde{F}(x - c, t + 1)] \quad (10.22)$$

and

$$\tilde{F}(x,t) = \sum_{i=1}^{3} u_i(x,t)V_i(x,t) \quad (10.23)$$

where $u_i(x,t)$ denotes the probability of using patch i:

$$u_i(x,t) = \frac{h(C_i(x,t))}{\Sigma_j h(C_j(x,t))} \quad (10.24)$$

As in dynamic programming, there is a stationary solution:

$$u_i(x) = u_i(x,t) \text{ for } t \ll T \quad (10.25)$$

Henceforth, we use this solution. From here on, the algorithm for finding the ESS is the same as before. First, we obtain the transition probabilities $M_{xy} = \Pr\{X(t+1) = y | X(t) = x\}$. Next, we use forward iteration, eq. 10.9, to calculate the stationary state distribution π_x. Finally, eq. 10.10 gives us N_i = average number of individuals using patch i, so that

$$f_2' = \frac{N_2}{N_2 + N_3}$$

as before. This specifies the response function $f_2' = R(f_2)$ for the patch-selection model with errors.

Figure 10.3 shows the response functions for three different values of the cost coefficient $c_0 = 0.0001, 0.001,$ and 0.01. For $c_0 = 0.0001$ (errors are mostly avoided) the response curve is a smoothed-out version of the original curve without errors (fig. 10.1), and the ESS remains dynamically unstable. As c_0 is increased (larger errors are tolerated), the response curve levels out, and the ESS stabilizes. (Note in fig. 10.3 that the evolutionary stable value $\bar{f}_2$ decreases as the parameter c_0 is increased. Can you explain this result intuitively?) Therefore, we can add a fourth way in which predator–prey oscillations could be stabilized, to the three ways suggested by Schwinning and Rosenzweig (1990), namely, the prey make imperfect habitat choices.

The paper by McNamara and Houston (1990) on state-dependent ideal free distributions uses a model similar to ours. Their calculations result in a smooth response curve, as a result of assuming stochastic variation in food sizes. This is often a useful procedure for smoothing the predictions of a dynamic state variable model (Houston and McNamara 1999, p. 35). Such stochasticity doubtlessly often reflects natural situations, although to include it in an empirically based model would add to the difficulties of parameter estimation.

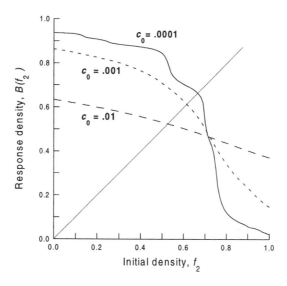

Figure 10.3 Response curves for the patch-selection model with decision errors for three different values of c_0.

This analysis could be extended in several ways. What happens when there are more than two foraging patches? What if the food recovery rate depends on the number of foragers on a patch? What if predator behavior is modeled explicitly?

The method described in this section—backward iteration followed by forward iteration to obtain the population-level response, the whole process then iterated to calculate the ESS—was used by Lucas et al. (1996) to model caller versus satellite (noncaller) strategies in anurans. Their model includes state dependence, frequency dependence, and density dependence, facilitating analysis of the relative importance of these conditions and their interactions. Where the response-function iterations failed to converge, these authors reported only the result of the final iteration. Houston and McNamara (1987) encountered similar problems in a dynamic game model of bird singing.

Lucas et al. (1995) used the same approach to model cooperative breeding; in this work the iterations always converged. Crowley and Hopper (1994) also used the iteration method to study cannibalism as a density-dependent dynamic game.

A final remark concerning forward iteration. Here, we have used the Markov chain method, which as we pointed out gives the exact population distribution of the state variable $X(t)$. Assuming that the model itself is stationary (meaning that the model parameters are constant over time), these distributions π_t converge to a stationary distribution $\bar{\pi}$, which then results in a stationary response function R. To compute the ESS, we iterate this procedure, hoping to obtain convergence.

An alternative to the Markov chain method would be Monte Carlo forward iteration, as elsewhere in this book. The only problem is that these simulated distributions π_t' will not converge strictly to a stationary distribution. Therefore it would be necessary to treat the distribution π_t' for moderately large t, as an approximately stationary distribution. One should inspect several successive distributions π_t' to see whether they seem approximately stationary; if so, one could average them to obtain a better approximation. Then, the resulting response function R will also be an approximation, so that the sequence $f', f'', \ldots$ of population-level responses will also be approximate. It may be hard to tell whether this sequence is (approximately) converging. We encourage you to try this approach on the patch-selection model. In most cases, a graphical plot of the iterates $f', f'', \ldots$ should indicate whether a reasonably constant "limiting" value $\bar{f}$ emerges. If not, the possibility of assuming errors in decisions could be investigated.

10.3 Genetic algorithms as optimization tools

Genetic algorithms and optimization in biology

Genetic algorithms are methods for solving optimization problems by mimicking natural selection (Holland 1975; Davis 1987, 1991; Goldberg 1989; Sumida et al. 1990; Beasley et al. 1993a,b). They are one technique in a family of recent methods in computer science that have drawn ideas from biology (Aarts and Korst 1989; Bishop 1995). Although genetic algorithms are often used to solve optimization problems that have no connection to biology, nothing prevents us from using them in a biological situation (Sumida et al. 1990). Of course, it should be recognized that the genetic algorithm is merely a computational technique and is not designed to represent the actual genetics of the population in any way. Thus, terms such as *genotype* and *evolution* are used in a purely descriptive sense, referring to the computation, not to the underlying biology. This comment also applies to the amount of variation from optimality predicted by the genetic algorithm—this is numerical variation and not bona fide genetic variability. Nevertheless, it is intellectually satisfying to have a method that predicts a certain amount of persistent variation in biological optimization models.

Genetic algorithms are particularly appropriate for optimization problems that involve state spaces that are simply too enormous for stochastic dynamic programming to provide an efficient solution. For example, if there is a high level of environmental uncertainty and unpredictability such as a pelagic larva might encounter with highly variable oceanic currents (Huse and Giske 1998; Giske et al. 1998), then it is difficult to store all of the environmental states for stochastic dynamic programming. Genetic algorithms provide a means for studying the evolution of behavior and the force of selection on behavior. Second, genetic algorithms necessarily work at the population level, rather than at the individual level alone. Because of this, genetic algorithm

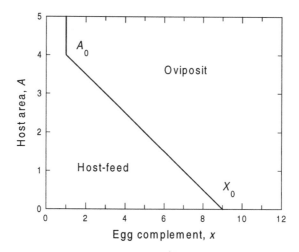

Figure 10.4 When a parasitoid can use hosts either for host-feeding or oviposition, the field data (see fig. 4.14) suggests that host-feeding occurs either when the egg complement is low or the host area is small. The roles of A_0 and X_0 are explained in the text.

models can readily be designed to incorporate realistic biological details such as density and frequency dependence, coevolution, and so on.

Using genetic algorithms

To use genetic algorithms, we require three kinds of submodels that address the following questions:

1. How do "genes" affect behavior (e.g., Plomin et al. 1990)?
2. How is reproductive success of different genes evaluated?
3. How is genetic information transmitted?

As with real organisms, the transmission of genetic information in computer organisms can involve either fundamentally continuous traits (i.e., quantitative genetics based on large numbers of loci each contributing a small effect) or discrete traits with individual loci. In the example that we consider in this section, we use mainly continuous traits for the computations, but we discuss models based on discrete traits at the end of the section.

An example: host-feeding by insect parasitoids

To help fix ideas and illustrate how genetic algorithms are used, we return to the end of chapter 4. There, we discussed a field study of an *Aphytis* parasitoid that can either feed on hosts or oviposit in them. The field data (also see Heimpel et al. 1998) and our knowledge of the predictions of other models suggest (fig. 10.4) that we can divide the egg complement/host area plane into regions in which host-feeding is the optimal behavior or in which oviposition is the optimal behavior. It would be possible, of course, to develop a dynamic state variable model to predict this division, and Heimpel et al. (1998) did this, but the dynamic state variable model is quite complicated even for a simple case of only three host sizes. Here, we explain how a genetic algorithm could be used to solve the same problem.

A quantitative genetic algorithm

To begin, we will use a quantitative genetic algorithm. The first question is how the genotype is connected to behavior. We assume that the genotype G_i of the ith individual consists of a pair of numbers $\{A_{0i}, X_{0i}\}$, where A_{0i} is the host area that corresponds to the boundary curve at $X = 1$ egg (see fig. 10.4) and X_{0i} is the value of the egg complement at which the boundary curve intersects the x-axis. Given the genotype, we characterize behavior by computing the boundary curve

$$S_i(x) = A_{0i} - \frac{A_{0i}}{X_{0i} - 1}(x - 1), \ x \le X_{0i} \qquad (10.26)$$

and assume that when a host is encountered and the egg complement of the parasitoid is x, the host is used for oviposition if the area is greater than $S_i(x)$ and the host is used for feeding otherwise. Equation 10.26 relates the genotype G_i to behavior.

Next, we conduct a Monte Carlo forward iteration to evaluate the actual reproductive success of each individual in the population. To do this, one needs to initialize the genotype distribution of the population. For the results that follow, in the first generation, we assumed that A_{0i} is normally distributed with mean m_a and standard deviation σ_a, that X_{0i} is normally distributed with mean m_x and standard deviation σ_x, and that the initial egg complement is uniformly distributed between one and the maximum egg complement x_{max} (table 10.2). We used the Box–Mueller algorithm (Press et al. 1986) to generate normally distributed random variables. Finally, since we are studying a host-feeding problem, we also need to characterize the initial level of reserves Y_i of the ith parasitoid; this too is assumed to be uniformly distributed between zero and its maximum y_{max}.

Reproductive success is computed in the following manner. For each individual in the population, we cycle for time t between $t = 1$ and the time horizon T. In each period, we determine whether or not a host is encountered and, if it is encountered, whether it is used for host feeding or oviposition. Doing this requires a description of the host environment. We assume that there are H different types of hosts, that λ_0 is the probability that no host is encountered in a single period, and that the probability of encountering host type i is given by

$$\lambda_i = \frac{1 - \lambda_0}{H} \ (i = 1, 2, \ldots, H) \qquad (10.27)$$

Table 10.2 Parameters used in the genetic algorithm

Parameter	Interpretation	Value
m_a	Mean of A_{0i}	1.5
m_x	Mean of X_{0i}	5.0
σ_a	Standard deviation of A_{0i}	0.6
σ_x	Standard deviation of X_{0i}	2.0
$x_{\max}$	Maximum egg complement	10
$y_{\max}$	Maximum amount of reserves measured in eggs	8
T	Time horizon for an individual forager	120
H	Number of different kinds of hosts	5
f	Fraction of reserves matured into eggs in one period	0.2
m_0	Mortality per period if no host is encountered	0.01
m_f	Mortality if the parasitoid host-feeds	0.03
m_{ov}	Mortality if the parasitoid oviposits	0.02
h^2	Heritability	0.65

If host type i is used for oviposition, the probability that an offspring emerges is given by

$$\phi_i = 0.6 \left(\frac{i}{H} \right)^3 \tag{10.28}$$

and if it is used for host-feeding, reserves are increased by

$$h_i = 1.5 \frac{i}{H} \tag{10.29}$$

The choices in eqs. 10.27–10.29 are arbitrary, intended to illustrate how genetic algorithms work. Heimpel et al. (1996, 1998) give details on empirical measures of ϕ_i and h_i; our choices are reasonable in the light of their data.

As with a dynamic state variable model, the genetic algorithm used here requires states; these are the egg complement $X_i(t)$ and the reserve level $Y_i(t)$ of the ith parasitoid at the start of period t. Reserves are converted to eggs at a fraction f per period. Thus, if no host is encountered, the state dynamics are given by

$$X_i(t+1) = X_i(t) + fY_i(t)$$
$$Y_i(t+1) = (1-f)Y_i(t) \tag{10.30}$$

and the parasitoid survives to period $t+1$ with probability e^{-m_0}. If a host is encountered and used for host-feeding, the state dynamics are given by

$$X_i(t+1) = X_i(t) + fY_i(t)$$
$$Y_i(t+1) = (1-f)Y_i(t) + h_i \tag{10.31}$$

and the parasitoid survives with probability e^{-m_f}. If the host is used for oviposition, the state dynamics are given by

$$X_i(t+1) = X_i(t) + fY_i(t) - 1$$
$$Y_i(t+1) = (1-f)Y_i(t) \tag{10.32}$$

and the parasitoid survives with probability $e^{-m_{ov}}$. As with a dynamic state variable model, the state dynamics are constrained by critical values (zero) and maximum values (x_{max}, y_{max}).

When a host is used for oviposition, the reproductive success $R_i(t)$, of the parasitoid is increased, provided that an offspring emerges. Thus,

$$R_i(t+1) = \begin{cases} R_i(t) + 1 & \text{with probability } \phi_i \\ R_i(t) & \text{with probability } 1 - \phi_i \end{cases} \tag{10.33}$$

As usual, the Monte Carlo forward iteration uses the random number generator to make this determination. Similarly, we determine whether the parasitoid survives period t by generating another random variable.

After cycling through all of the N parasitoids in the population, we obtain the total reproductive success of the population:

$$R_{\text{pop}} = \sum_{i=1}^{N} R_i(T) \tag{10.34}$$

As discussed here earlier, it is best to assume that the population is at a stable level, in the sense that R_{pop} is approximately equal to N (there will, of course, be fluctuations in population size from one year to the next, even for stable populations). Thus, if R_{pop} exceeds N, we assume that a fraction N/R_{pop} of the offspring survives to become reproductive in the next generation. This fraction is also determined by comparing a uniformly distributed random variable with the smaller of 1 or N/R_{pop}.

In general, a mother that has genotype $\{A_{0i}, X_{0i}\}$ may produce j offspring, $j = 0, 1, \dots$. Suppose that $\{A_{ij}, X_{0ij}\}$ is the genotype of the jth offspring of the ith mother. It is here that the genetics arises. We assume that the offspring's genotype is determined by a combination of maternal and environmental effects, weighted according to the heritability (Falconer 1993)

$$A_{ij} = h^2 A_i + (1 - h^2)(m_a + \sigma_a E_a)$$
$$X_{0ij} = h^2 X_{0i} + (1 - h^2)(m_x + \sigma_x E_x) \tag{10.35}$$

where E_a and E_x are normally distributed random variables with mean zero and variance one. Equation 10.35 is best understood in the limiting cases. If $h^2 = 1$, then offspring are identical to their mothers. If $h^2 = 0$, then offspring traits are drawn from a normal distribution, independent of the mother's traits. For $0 < h^2 < 1$, offspring traits are a mixture of the mother's traits and a variable drawn from a normal distribution that characterizes the variation in traits in the population.

Results for the quantitative genetic algorithm

As the program described in eqs. 10.26–10.35 is run, the effect of initial conditions weakens, and evolution of the genotype distribution occurs, determined by reproductive success and genetic transmission. Here, we report results following initialization and after 500 generations of evolution.

Typical initial distributions of A_0 and X_0 are shown in fig. 10.5a,b. In the initial generation, there is little relationship between the frequency of a genotype and its expected reproductive success (fig. 10.5c).

After the genetic algorithm is run for many generations, a different picture emerges (fig. 10.6). First, note that the marginal distributions of A_0 and X_0 have become narrower (that of X_0 considerably more so than that of A_0). Most importantly, note that there is an extremely close relationship now between the frequency of a genotype in the population and its expected reproductive success. However, the population is by no means monomorphic: many different genotypes are present. Thus, one would expect variability in a population of insects following host-feeding and oviposition based on the rules that we derived here.

Interesting trade-offs can be identified (although we will not pursue them in detail). The rate of evolution in the genetic algorithm is determined by the heritability h^2 and the genetic variances σ_a^2 and σ_x^2. When heritability is larger and genetic variances are smaller, individuals reproduce more truly to themselves. This slows the rate of evolution toward the optimal solution. On the other hand, when heritability is smaller and genetic variances are larger, both the rate of evolution and the production of spurious mutations are higher, thus leading toward the representation of many different kinds of genotypes in the population. Indeed, one of the challenges in using genetic algorithms to solve optimization problems is to determine the best mix of heritability and genetic variance. In biology, the problem is different because we can determine these quantities from measurements on parents and their offspring (Falconer 1993).

A discrete genetic algorithm

In the quantitative genetic algorithm, we allowed A_0 and X_0 to be continuous variables, which is equivalent to assuming that a very large number of loci each contribute a small amount to the determination of A_0 or X_0. A discrete genetic algorithm considers that A_0 and X_0 are determined by a relatively small number of loci. Here, we might imagine that each locus has two alleles, which could be labeled zero or one. For simplicity, we assume that each gene has m loci. Thus, now the genotype of the ith individual is $G_i = \{a_{0i}, x_{0i}\}$, but a_{0i} and x_{0i} have representations as $a_{0i} = \{a_{0i1}, a_{0i2}, \ldots a_{0im}\}$ and

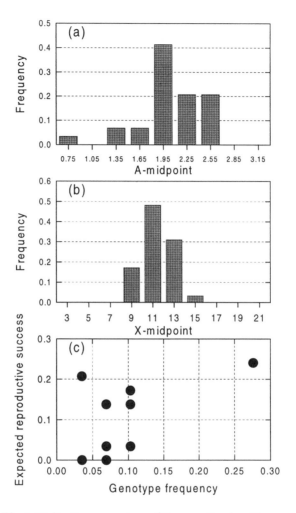

Figure 10.5 The initialization procedure of the genetic algorithms produces distributions of A_0 (panel a) and X_0 (panel b) that are peaked. We chose increments of 0.3 for A_0 and 2.0 for X_0. Note that little relationship exists between the initial frequency of a genotype in the population and its expected reproductive success (panel c), as would be expected when the genotype is randomly chosen. (A linear regression is driven by the outlier at relatively high frequency).

$x_{0i} = \{x_{0i1}, x_{0i2}, \ldots x_{0im}\}$. Because there are two alleles at each locus, the values that a_{0ik} and x_{0ik} can take are 0, 1, and 2 (homozygote, heterozygote, homozygote, respectively). From these representations, we construct A_{0i} and X_{0i}. For example, we might set

$$A_{0i} = \left\{ \frac{1}{2m} \sum_{k=1}^{m} a_{0ik} \right\} a_{\max} \qquad (10.36)$$

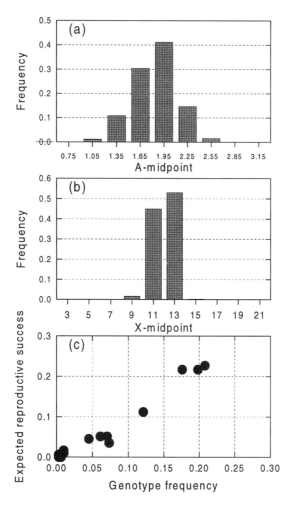

Figure 10.6 The same distributions as were plotted in fig. 10.5, but after 500 generations of evolution by the genetic algorithm. Note the tightening of all distributions and the strong relationship between the frequency of a genotype in the population and its expected reproductive success.

where a_{max} is the maximum value of the area of hosts encountered. A_{0i} constructed in this manner will vary from zero to a_{max}.

There is more leeway in constructing X_{0i} from the genotype. In principle, X_{0i} can vary between zero and ∞, so we might use a construction such as

$$X_{0i} = 1 + \frac{\sum_{k=1}^{m} x_{0ik}}{2m - \sum_{k=1}^{m} x_{0ik}} \tag{10.37}$$

The construction of the boundary curve, evaluation of reproductive success, and number of offspring proceeds as with the quantitative genetic algorithm.

The transmission of genetic information can be much richer with the discrete genetic algorithm. We implicitly follow females, so we need to make an assumption about the distribution of male genotypes in the population and if there is any kind of assortative mating. We can also include known biological effects such as crossover, epistatis, etc. For the current problem, it is not clear that the discrete genetic algorithm would add much to our understanding.

Recent papers that use genetic algorithms in biology include the following. Price (1994) studied the behavioral ecology of begging in yellow-headed blackbird (*Xanthocephalus xanthocephalus*) nestlings. Barta et al. (1997) considered a scrounger-forager model with spatial inhomogeneity. Robertson et al. (1998) modeled natural tolerance and communal associations among related females. Hoffmeister and Roitberg (1998) studied pheromone persistence in parasitoids.

11

Stochasticity, Uncertainty, and Information as a State Variable

Until now, we have assumed that the world is stochastic: various events such as finding food or avoiding predators are characterized by probabilities or probability distributions. There is little that can be done to control stochastic events since they are probabilistic by their nature. However, the world may also be uncertain: the parameters that characterize the probability of finding food or avoiding predation may themselves have probability distributions that characterize imperfect *knowledge*, rather than uncertain *events*. Learning may reduce the uncertainty that results from imperfect knowledge, as in the model of functional response discussed in chapter 9.

To model learning, information must be incorporated as a state variable. There is a rich literature on this subject, including Krebs et al. (1978), Mangel and Clark (1983), Clark and Mangel (1984), Mangel and Roitberg (1989), Mangel (1990c), and Bouskila and Blumstein (1992). There are many subtleties and nonintuitive aspects of informational problems. In the next section, we use a simple model of ephemeral patches to illustrate how learning can be characterized and used in state variable models.

11.1 Ephemeral patches and uncertain food

To illustrate ideas about information and learning, we first consider a model for ephemeral patches. In particular, imagine two patches that emerge at time $t = 1$ and disappear at time $t = T$; food will be only in one of these patches, with probability p_1 that it is in patch 1 and $p_2 = 1 - p_1$ that it is in patch 2. Examples are a predator looking for a prey animal that has fled into cover, a golfer looking for his ball, and a search team looking for a lost hiker. A forager searches patch i and, given that the food is in patch i, finds the food in a single period with probability $p_d(i)$. We want to predict the optimal

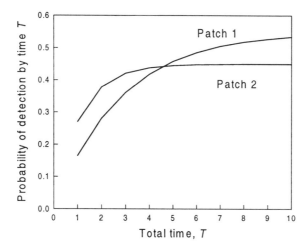

Figure 11.1 The fixed-patch strategy has the organism starting and staying in one patch, determined according to which of $V_{\text{fixed}}(1)$ or $V_{\text{fixed}}(2)$ is higher. Here we show results for $p_1 = .55$, $p_d(1) = .3$ and $p_d(2) = .6$. When $T = 10$, the best patch to visit is patch 1; however, if $T < 5$, the best patch to visit is patch 2.

patch visitation strategy for the forager. To do this, we will investigate a series of models for behavior.

Model 1. Stay in one patch

First, we suppose that the forager stays in one patch for the entire time $t = 1$ to $t = T$. Which patch should the forager pick? Common sense suggests that a good rule should involve the probability of detecting food by time T; we denote these probabilities by $V_{\text{fixed}}(i)$ for $i = 1$ or 2. To evaluate them, note that the probability that the forager does not detect food in a single period of search in patch i, given that food is in patch i, is $[1 - p_d(i)]$. Hence, the probability that it does not detect food in T periods is $[1 - p_d(i)]^T$ and the probability that it detects food is $1 - [1 - p_d(i)]^T$. Combining these with the probability that food is in the patch leads to

$$V_{\text{fixed}}(i) = p_i\{1 - [1 - p_d(i)]^T\} \tag{11.1}$$

Then we predict that the forager will go to the patch with the highest value of $V_{\text{fixed}}(i)$. Note that for $T = 1$,

$$V_{\text{fixed}}(i) = p_i \cdot p_d(i)$$

which is the probability of finding food in patch i in a single attempt. For example, suppose that

$$p_1 = .55, \quad p_d(1) = .3, \quad p_d(2) = .6$$

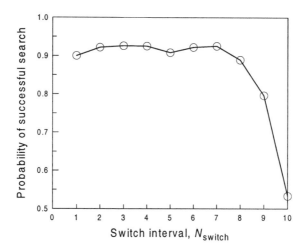

Figure 11.2 The probability of successful search is relatively constant for values of $N_{\text{switch}} < 7$ or so, and above that it begins to drop. For values less than 7, the small differences in probability are due to the limited number of simulations.

Then for $T = 1$,

$$V_{\text{fixed}}(1) = 0.165, \text{ and } V_{\text{fixed}}(2) = 0.270$$

so that patch 2 should be searched—even though the food is more likely to be in patch 1.

Figure 11.1 shows $V_{\text{fixed}}(i)$ as a function of the number of search periods T. For $T < 5$, it is best to search patch 2, but for larger values of T, the best fixed-patch strategy is to search patch 1. Note that from eq. 11.1, $V_{\text{fixed}}(i)$ approaches p_i, as T increases: with many opportunities to search the patch, the probability of success in patch i approaches p_i. Thus patch 1 should be searched if T is large, but not if T is small, in this example.

Model 2. Switch at fixed intervals

The fixed-patch strategy makes no use of information obtained while searching. Common sense suggests, however, that as time progresses and the search is unsuccessful, food is more likely to be in the other patch. Thus, we might envision a forager that periodically switches between patches, as long as the search is not successful. Assume that the forager switches patches after N_{switch} unsuccessful searches in a patch. We used Monte Carlo simulation (forward iteration) to calculate the probability of success in $T = 10$ periods, as a function of N_{switch}, assuming that the forager begins in patch 1—see fig. 11.2. We see that the probability of successful search is relatively constant for values of $N_{\text{switch}} < 7$ or so, and after that it begins to drop. The small differences for values less than seven are sampling fluctuations due to a limited number of simulations. A general rule of thumb is that fluctuations in simulation results scale as $1/\sqrt{n}$, where n is the number of simulations. Here we simulated 10,000 foragers, so that fluctuations of the order of $1/\sqrt{10000} = 0.01$ should be anticipated. With this reasoning, we consider that N_{switch} ranging from two to six all give the essentially the same result, that the probability of success is about .92. This is much superior to the fixed behavior strategy (.53), and we would expect selection pressure for search with a switching rule.

Bayesian Updating and Learning

By far the most logically consistent method for incorporating probabilistic information is that of Bayesian analysis; see the appendix to this chapter; also see Berger and Berry (1988), Hilborn and Mangel (1997), Howson and Urbach (1989, 1991), Baldi and Brunach (1998). Suppose that we let

$$p(n_1, n_2) = \Pr \left\{ \begin{array}{l} \text{the food is in patch 1, given } n_1 \text{ unsuccessful searches} \\ \text{in patch 1 and } n_2 \text{ unsuccessful searches in patch 2} \end{array} \right\} \quad (11.2)$$

In Bayesian analysis, $p(n_1, n_2)$ is called the *posterior probability* that the food is in patch 1, given the information of unsuccessful search. Before any search, $p(0,0) = p_1$; in Bayesian analysis this is called the *prior probability* that the food is in patch 1. Once the search has commenced, we evaluate $p(n_1, n_2)$ by applying Bayes' theorem (see appendix):

$$\Pr \left\{ \begin{array}{l} \text{the food is in patch 1, given } n_1 \text{ unsuccessful searches} \\ \text{in patch 1 and } n_2 \text{ unsuccessful searches in patch 2} \end{array} \right\}$$

$$= \frac{\Pr \left\{ \begin{array}{l} \text{the food is in patch 1 and } n_1 \text{ unsuccessful searches} \\ \text{in patch 1 and } n_2 \text{ unsuccessful searches in patch 2} \end{array} \right\}}{\Pr \left\{ \begin{array}{l} n_1 \text{ unsuccessful searches in patch 1 and } n_2 \\ \text{unsuccessful searches in patch 2} \end{array} \right\}} \quad (11.3)$$

The probability that the food is in patch 1 and there are n_1 unsuccessful searches in patch 1 is $p_1[1 - p_d(1)]^{n_1}$. Similarly, the probability that the food is not found after n_1 searches in patch 1 and n_2 searches in patch 2 is $p_1[1 - p_d(1)]^{n_1} + (1 - p_1)[1 - p_d(2)]^{n_2}$. Hence we conclude that

$$p(n_1, n_2) = \frac{p_1[1 - p_d(1)]^{n_1}}{p_1[1 - p_d(1)]^{n_1} + (1 - p_1)[1 - p_d(2)]^{n_2}} \quad (11.4)$$

Since we assume search in every time period, we set $n_1 + n_2 = t$ where t is the total number of periods of searching. In fig. 11.3, we show $p(n_1, n_2)$ for $t = 4, 6$, and 8 when $p_d(1) = .3$, $p_d(2) = .6$ and $p_1 = .55$. For example, suppose that $t = 4$ and $n_1 = 0$. Then $n_2 = 4$—there were 4 unsuccessful searches in patch 2. Thus food is most likely to be in patch 1; in fact $p(0, 4) = .979$.

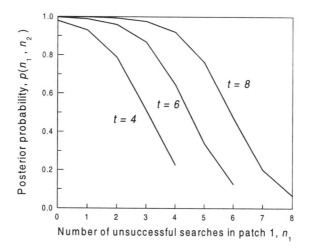

Figure 11.3 The posterior probability $p(n_1, n_2)$ that the food is in patch 1, given n_1 unsuccessful searches in patch 1 up to times $t = 4$, 6, or 8 for the parameters used above.

Model 3. Search the most likely patch

Now suppose that the forager uses the Bayesian updating formula 11.4 to calculate the posterior probability $p(n_1, n_2)$ that the food is in patch 1. Then, a reasonable ("myopic") search rule is always to search the patch that has the higher probability—that is, search patch 1 if $p(n_1, n_2) > .5$ and vice versa (if $p(n_1, n_2) = .5$ then it is immaterial which patch is searched next). Here we can consider n_1 as an **informational state variable**. The search decision at each time t is determined by the current informational state.

Again we used Monte Carlo forward iteration to calculate the probability of successful search using the Bayesian method and rule. With the same parameter values as used in the fixed switch-time rule, we obtained probability of success $= .923$, which is essentially the same as the fixed switching rule. An investigation of other parameter values (table 11.1) suggests that the conclusion is robust: here there is little advantage of sophisticated learning over simple switching.

Model 4. The trade-off between information and mortality

The previous results show that by dealing with the information provided by unsuccessful search—either by using a fixed counting rule for switching or a Bayesian rule—the forager can greatly improve its chance of finding food before the patches disappear. However, there may be costs associated with moving from one patch to the other. We can capture these costs by assuming that the forager faces a mortality risk M when moving between patches. Then, when comparing the rules that involve information processing with the one that has the organism stay in a patch, we discount the probability of success

Table 11.1 Probability of success, for different parameter values, of the three behavioral rules

Parameters	Probability of success according to:		
	Stay in one patch	Fixed switching[a]	Bayesian
$p_1 = .5$, $p_d(1) = p_d(2) = .3$	.48	.83 (1–6)	.83
$p_1 = .5$, $p_d(1) = p_d(2) = .2$	.44	.67 (1–6)	.67
$p_1 = .5$, $p_d(1) = p_d(2) = .1$	.33	.42 (1–5)	.41

a. We show the highest value of the success probability over different values of the switching rule and in parentheses, the switching rules that are within .01 of the highest probability of success.

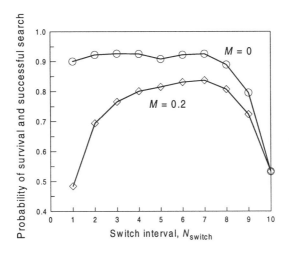

Figure 11.4 The probability of surviving and finding food, for the fixed switch strategy ($M = 0.2$). The case $M = 0$ is the same as in fig. 11.2.

by $\exp(-MN_{\text{moves}})$, where N_{moves} is the number of moves associated with that switching rule. We can interpret this result as the probability of finding food and surviving the moves; fig. 11.4 shows the results for the fixed switching rule. Notice that a low value of N_{switch} (which would imply frequent switching) is undesirable now.

Model 5. A dynamic state variable model

When there is mortality associated with staying in a patch or moving between patches, we would actually like to compute optimal behavioral rules that take mortality into account during the development of the rules, rather than after the fact as we have done. This requires constructing a dynamic state variable model with information as the state variable, and information updating as part of the dynamics of the state variable.

To do this, we let

$$F(n, j, t) = \text{maximum probability that the forager finds} \qquad (11.5)$$
$$\text{the food and survives between } t \text{ and } T, \text{ given}$$
$$\text{that it has had } n \text{ unsuccessful searches in}$$
$$\text{patch 1 and currently is in patch } j$$

Assuming that the search terminates at the start of period T, the end condition is given by

$$F(n, 1, T) = F(n, 2, T) = 0 \text{ for every value of } n \qquad (11.6)$$

For previous periods, the forager can either stay in the current patch or move to the other patch. If the fitness values of these options are $V_{\text{stay}}(n, j, t)$, and $V_{\text{move}}(n, j, t)$ the dynamic programming equation is

$$F(n, j, t) = \max\{V_{\text{stay}}(n, j, t), \ V_{\text{move}}(n, j, t)\} \qquad (11.7)$$

To be specific, now we will evaluate the fitness of staying or moving for a forager that is currently in patch 1. If the forager stays,

$$V_{\text{stay}}(n, 1, t) = p(n_1, n_2)[p_d(1) + (1 - p_d(1))F(n + 1, 1, t + 1)]$$
$$+ (1 - p(n_1, n_2))F(n + 1, 1, t + 1) \qquad (11.8)$$

where $n_1 = n$ and $n_2 = t - 1 - n$. If food is in patch 1 (with probability $p(n_1, n_2)$), the forager may detect it or not. If food is in patch 2 (with probability $1 - p(n_1, n_2)$), there is no chance of finding it when the forager stays in patch 1. When the forager moves to patch 2, the roles of patch 1 and patch 2 are interchanged, and the cost of mortality is included. We obtain

$$V_{\text{move}}(n, 1, t) = e^{-M}\Big\{p(n_1, n_2)F(n, 2, t + 1)$$
$$+ \big(1 - p(n_1, n_2)\big)[p_d(2) + \big(1 - p_d(2)\big)F(n, 2, t + 1)]\Big\} \qquad (11.9)$$

Analogous equations can be derived for the fitness values of moving and staying when the forager is currently in patch 2.

As the mortality associated with movement increases, the advantage of following the optimal Bayesian behavior rather than the best fixed switch behavior increases and can be considerable when movement mortality is high (fig. 11.5). Note too that the fitness of the myopic Bayesian behavior (model 3) rapidly declines with increasing mortality, because a forager that follows the myopic Bayesian behavior will make many moves between patches.

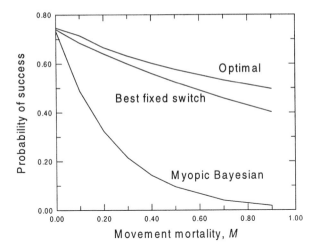

Figure 11.5 Comparison of the best fixed switch, myopic Bayesian, and optimal Bayesian behaviors for the probability of detecting food and avoiding predation as mortality increases, when the forager starts in patch 1. Here the probability of success is the probability of finding food and surviving.

11.2 A Bayesian foraging model

In this section we consider a somewhat different informational problem, in which the abundance of food in a certain habitat fluctuates from one year to the next. At the start of the year the forager does not know how abundant food is but uses the information obtained while foraging to estimate this parameter. Mangel and Clark (1983) considered a similar problem in fisheries, using a more complicated model.

We consider a forager with a single foraging habitat. The probability of encountering a food item (of constant size Y) during a single period interval is λ. The predation risk per period is $\mu(x)$, where x denotes the forager's energy reserves, $0 \leq x \leq x_{\max}$. The problem is to find the foraging strategy that maximizes the probability of surviving the winter consisting of T periods. In each period, the forager's only decision is whether to forage or rest. The metabolic cost is c per period, whether foraging or resting.

So far, the problem resembles the patch-selection problem of chapter 1; starvation risk can be reduced by maintaining high energy reserves, but this increases the risk of predation while foraging. Here we make the additional assumption that the forager does not know the current year's value of λ at the beginning of the season but may be able to estimate λ from its success rate when foraging. Forbes and Ydenberg (1992) consider a related problem for optimal siblicide in osprey (*Pandion haliaetus*) nestlings that use Bayesian updating to infer the current season's food abundance from the rate at which their parents deliver food to the nest. Another example occurs in Roitberg (1990).

We will compare several models related to this problem.

1. Perfect information. The forager knows the value λ at the beginning of winter.
2. No use of current information. The forager uses the average value $\overline{\lambda}$ throughout the season.
3. Bayesian updating. The forager estimates the current season's value of λ as a result of its foraging success.

By comparing the initial fitness of a forager in these three cases, we can assess the relative importance of using Bayesian methods to estimate current environmental parameters.

We assume that λ fluctuates from year to year. For simplicity, we assume that λ has a discrete prior distribution: $p_j = \Pr\{\lambda = \lambda_j\}$. For the numerical calculations we take

$$\lambda_j = .2, .4, .6, \qquad p_j = 1/3, 1/3, 1/3$$
$$Y = 3, \quad c = 1, \quad x_{\max} = 30, \quad T = 60$$
$$\mu(x) = .01 + \frac{.05x}{x_{\max}}$$

where $\mu(x)$ denotes the predation risk per period.

Model 1. Perfect information

Here we assume that the forager knows the value of λ for the current year. We let $F_P(x, t; \lambda)$ be the fitness function (survival) under this assumption ("P" for perfect). Then

$$F_P(x, T; \lambda) = \phi(x) \tag{11.10}$$

where $\phi(0) = 0$ and $\phi(x) = 1$ for $x > 0$. Also

$$F_P(x, t; \lambda) = \max\{V_0, V_1\} \tag{11.11}$$

where V_0 and V_1 correspond to resting or foraging, respectively:

$$V_0 = F_P(x - c, t + 1; \lambda) \tag{11.12}$$
$$V_1 = [1 - \mu(x)][\lambda F_P(x + Y - c, t + 1; \lambda) + (1 - \lambda) F_P(x - c, t + 1; \lambda)] \tag{11.13}$$

The optimal strategy is a threshold strategy:

$$\text{forage in period } t \text{ if and only if } x(t) < x^*(t; \lambda) \tag{11.14}$$

In other words, whenever its reserves are too low, the forager should forage, attempting to build reserves up to the level $x^*(t; \lambda)$. Figure 11.6 shows the optimal threshold $x^*(t; \lambda)$ for three values of λ. (Can you explain these curves? Why does a smaller λ lead to higher thresholds? Compare with fig. 5.12.)

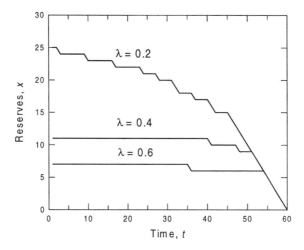

Figure 11.6 Threshold reserve levels $x^*(t; \lambda)$ for the foraging model under perfect information about λ. The forager should forage if and only if $x(t) < x^*(t; \lambda)$. In the model, fitness is defined in terms of survival to time $T = 60$.

Taking into account the annual fluctuations in λ, where $\lambda = \lambda_j$ with probability p_j, the expected fitness of the optimal forager prior to the start of the year is given by

$$E_\lambda\{F_P(x_1, 1; \lambda)\} = \sum_j p_j F_P(x_1, 1; \lambda_j) \tag{11.15}$$

where x_1 is the initial reserve level, assumed given. Taking $x_1 = 5$, the numerical computation gives

$$E_\lambda\{F_P(x_1, 1; \lambda)\} = 0.242 \tag{11.16}$$

Model 2. No updating

Next we consider the case in which the forager does not know the current value of λ but does know the mean value $\overline{\lambda}$. Here, we assume that the forager does not use Bayesian updating to improve its estimate of the current value of λ. Thus, the best it can do is to behave as if $\lambda = \overline{\lambda}$, that is, the forager uses the threshold strategy $x^*(t; \overline{\lambda})$, as calculated from Model 1. To calculate the expected fitness of this strategy, we use the method described in section 10.2 for assessing errors in decisions. Thus, we define $F_N(x, t; \lambda)$ as the fitness of a forager using the threshold strategy $x^*(t; \overline{\lambda})$, when the current environmental parameter equals λ. Then,

$$F_N(x, T; \lambda) = \phi(x) \tag{11.17}$$

and

$$F_N(x, t; \lambda) = \begin{cases} [1 - \mu(x)][\lambda F_N(x + Y - c, t + 1; \lambda) \\ \quad + (1 - \lambda)F_N(x - c, t + 1; \lambda)] & \text{if } x < x^*(t; \overline{\lambda}) \\ F_N(x - c, t + 1; \lambda) & \text{if } x \geq x^*(t; \overline{\lambda}) \end{cases}$$
(11.18)

The average initial fitness is $E_\lambda\{F_N(x_1, 1; \lambda)\}$. Using the same numerical parameter values as in Model 1, we obtain

$$E_\lambda\{F_N(x_1, 1; \lambda)\} = 0.221$$
(11.19)

Comparing this value with that of eq. 11.16, we see that the use of perfect information about the current value of λ increases fitness by 9.5%, relative to the no-information case. We can interpret this increase in fitness as the "evolutionary value of perfect information."

We will not pause to undertake a detailed sensitivity analysis of this result. Qualitatively speaking, the more uncertain the environment (i.e., the more variable λ is), the greater the value of reducing this uncertainty. On the other hand, the more persistent environmental changes are, the more valuable it is to learn about them. In particular, if the encounter probability λ were to fluctuate rapidly, the information about the current value of λ would be of limited value (Stephens 1991). We discuss this point further in section 11.3.

Model 3. Bayesian updating

In reality, a forager would never possess perfect information about its environment. Rather, it would obtain information by experience. As discussed in section 11.1, this gradual accumulation of information can be modeled by treating information as a state variable and using the Bayesian updating formula. For our present foraging model, this requires two additional state variables:

$A(t) = $ number of foraging attempts made in periods $1, 2, \ldots, t - 1$

$S(t) = $ number of these attempts that were successful

Thus,

$$0 \leq S(t) \leq A(t) \leq t \leq T$$
(11.20)

The dynamics of these state variables are given by

$$A(t + 1) = \begin{cases} A(t) & \text{if no attempt is made in period } t \\ A(t) + 1 & \text{if an attempt is made in period } t \end{cases}$$
(11.21)

$$S(t + 1) = \begin{cases} S(t) & \text{if no encounter occurs in period } t \\ S(t) + 1 & \text{if an encounter occurs in period } t \end{cases}$$
(11.22)

Both the fitness function and the optimal strategy depend on these informational state variables, as well as on the current reserves $X(t)$. Thus,

$$F_B(x, a, s, t) = \text{maximum probability of surviving from } t \text{ to } T, \quad (11.23)$$
$$\text{given that } X(t) = x, \ A(t) = a, \ S(t) = s$$

Note that $F_B(x, a, s, t)$ is not a function of λ because the forager does not know λ. Instead, the forager uses Bayesian updating to estimate λ on the basis of its experience to date. Then (see appendix),

$$\Pr\{\lambda = \lambda_j | s \text{ encounters in } a \text{ attempts}\}$$

$$= \frac{\Pr\{s \text{ encounters in } a \text{ attempts} | \lambda = \lambda_j\} \Pr\{\lambda = \lambda_j\}}{\sum_k \Pr\{s \text{ encounters in } a \text{ attempts} | \lambda = \lambda_k\} \Pr\{\lambda = \lambda_k\}} \quad (11.24)$$

To calculate eq. 11.24, we use $\Pr\{\lambda = \lambda_j\} = p_j$ and

$$\Pr\{s \text{ encounters in } a \text{ attempts} | \lambda = \lambda_j\} = \binom{a}{s} \lambda_j^s (1 - \lambda_j)^{a-s}$$

that is, a binomial distribution. Hence, eq. 11.24 becomes

$$\Pr\{\lambda = \lambda_j | s \text{ encounters in } a \text{ attempts}\}$$

$$= \frac{\lambda_j^s (1 - \lambda_j)^{a-s} p_j}{\sum_k \lambda_k^s (1 - \lambda_k)^{a-s} p_k} = p_j'(a, s) \quad (11.25)$$

These probabilities $p_j' = p_j'(a, s)$ are the posterior probabilities of λ, given that s prey were encountered in a attempts. The posterior average of λ, given a and s, is expressed by

$$\overline{\lambda}(a, s) = \sum_j p_j'(a, s) \lambda_j \quad (11.26)$$

The dynamic programming equations are

$$F_B(x, a, s, T) = \phi(x) \quad (11.27)$$

$$F_B(x, a, s, t) = \max\{V_0, V_1\} \quad (11.28)$$

where

$$V_0 = F_B(x - c, a, s, t + 1) \quad (11.29)$$

$$V_1 = [1 - \mu(x)] \Big\{ \overline{\lambda}(a, s) F_B(x + Y - c, a + 1, s + 1, t + 1)$$

$$+ [1 - \overline{\lambda}(a, s)] F_B(x - c, a + 1, s, t + 1) \Big\} \quad (11.30)$$

Table 11.2 Initial fitness for Models 1, 2 and 3 (see text for parameter values)

Model	Initial fitness
1. Perfect information	0.2415
3. Bayesian updating	0.2331
2. No information	0.2211

This all looks straightforward, but there's a catch. Consider the size of the array $F_B(x, a, s, t)$, namely $(x_{max} + 1) \cdot T^3$ (because $0 \leq x \leq x_{max}$ and $0 \leq a, s \leq T$). With the parameters $x_{max} = 20$ and $T = 60$, the size of F_B is 4,536,000. If the values of F_B are calculated using double precision (requiring 8 bytes), this means that the array F_B requires 36.3 megabytes of CPU storage—beyond the capacity of some of today's PCs.

Storage requirements can be reduced somewhat, as follows. Rather than storing the entire array $F_B(x, a, s, t)$, we store only the values $F_B(x, a, s)$ for the last iteration, update these to $F_{new}(x, a, s)$, and then release the previous storage. This changes the memory requirement to $2(x_{max} + 1)T^2 = 151,200$, or 1.21 megabytes.

The point of this technical digression is to warn you that state variable models involving learning can rapidly become computationally unwieldy. Imagine, for example, trying to model learning about two environmental parameters in a single model. This would require at least two more state variables and would increase storage requirements several hundredfold. One wonders, how does the brain of an insect store and rapidly process several informational variables?

The initial fitness of the Bayesian forager is $F_B(x_1, 0, 0)$ for the final iteration ($t = 1$). Table 11.2 shows the initial fitness values for Models 1–3, for the assumed parameter values. Note that the Bayesian forager obtains fitness intermediate between the cases of zero information and perfect information.

11.3 Tracking a changing environment

Thus far, we have assumed that the parameter λ remains fixed over the entire time span $t = 1, 2, \ldots, T$ (though presumably it changes from one such span to the next). This represents an extreme case. The opposite extreme arises if λ varies unpredictably from each time period t to the next. In general, most environmental parameters have a certain tendency toward persistence and change more or less rapidly over time. A modified form of Bayesian updating can be used in this case.

To explain this, first we must reformulate the Bayesian updating algorithm used in the previous section, when λ remains constant. Now, rather than keeping track of $A(t)$, the number of attempts, and $S(t)$, the number of successes, suppose that the forager tracks the changing probabilities $p_j(t) = \Pr\{\lambda = \lambda_j\}$ given the search history to period t. These probabilities are updated from period t to $t+1$ by using Bayes' theorem in the following way:

$$p_j(t+1) = \begin{cases} p_j(t) & \text{if no attempt is made in period } t \\[2ex] \dfrac{\lambda_j p_j(t)}{\Sigma_k \lambda_k p_k(t)} & \text{if a successful attempt is made in period } t \\[2ex] \dfrac{(1-\lambda_j)p_j(t)}{\Sigma_k(1-\lambda_k)p_k(t)} & \text{if an unsuccessful attempt is made in period } t \end{cases}$$

$$(11.31)$$

These equations are special cases ($a = 0$ and $s = 0$, or $a = 1$ and $s = 1$ or 0) of eq. 11.25. This updating procedure is equivalent to the procedure obtained by using eq. 11.25 directly.

Now suppose that λ changes over time. Then recent foraging success is more representative of λ than earlier successes. One way to take account of this fact is to modify eq. 11.31 as follows:

$$p'_j(t+1) = \begin{cases} \alpha p'_j(t) + (1-\alpha)p_j(0) & \text{(no attempt)} \\[2ex] \dfrac{\alpha \lambda_j p'_j(t)}{\Sigma_k \lambda_k p'_k(t)} + (1-\alpha)p_j(0) & \text{(successful attempt)} \\[2ex] \dfrac{\alpha(1-\lambda_j)p'_j(t)}{\Sigma_k(1-\lambda_k)p'_k(t)} + (1-\alpha)p_j(0) & \text{(unsuccessful attempt)} \end{cases}$$

$$(11.32)$$

Here $p_j(0)$ are the prior probabilities, and α is a parameter with $0 \le \alpha \le 1$. If λ does not change over time, $\alpha = 1$ is the appropriate value. Here, all past experience is equally informative about the value of λ. In the general case of varying λ, the more rapidly λ varies, the less useful is information from the distant past. Using a value $\alpha < 1$ amounts to "discounting" information from period $t - i$ by a factor α^i. This question is discussed further by McNamara and Houston (1987); Mangel and Roitberg (1989); and Mangel (1990c).

11.4 Discussion

In this chapter we have indicated how learning about the environment can, in principle, be modeled using state variables. Our models considered only uncertainty in parameters related to foraging success, but other environmental variables can be handled in the same way. For example, a forager may need to estimate the current predation risk before deciding which patch to exploit, or how intensively to forage in a given patch. Many organisms employ inducible defenses in response to cues of predation risk (Harvell 1990). Some social insects convey information about the environment to colony members

(Oster and Wilson 1978). State variable models can help to increase our understanding of such adaptations (Clark and Harvell 1992).

Appendix: Bayesian updating

Bayes' formula is

$$\Pr\{A|B\} = \frac{\Pr\{B|A\}\Pr\{A\}}{\Pr\{B\}} \qquad (A.1)$$

This formula is sometimes called the "Law of Inverse Probability." To derive Bayes' formula, we need recall only the definition of conditional probability, eq. 1.6:

$$\Pr\{A|B\} = \Pr\{A,B\}/\Pr\{B\}$$

$$\Pr\{B|A\} = \Pr\{B,A\}/\Pr\{A\}$$

Since $\Pr\{A,B\} = \Pr\{B,A\}$, eq. A.1 follows by combining these two equations. It is interesting that such a simple mathematical result has such profound implications!

In practice, Bayes' formula is usually used in a special form. Consider a number of events $A_1, A_2, \ldots, A_n$, which are mutually exclusive (A_i and A_j can't both occur in the same experiment) and exhaustive (one of the A_i must occur in any instance of the experiment). In symbols, $\Pr\{A_i, A_j\} = 0$, $(i \neq j)$, and $\sum_{i=1}^{n} \Pr\{A_i\} = 1$. Then we have

$$\Pr\{B\} = \sum_{j} \Pr\{B, A_j\}$$

$$= \sum_{j} \Pr\{B|A_j\}\Pr\{A_j\}$$

by eq. 1.6. Hence Bayes's formula A.1 can be written as

$$\Pr\{A_i|B\} = \frac{\Pr\{B|A_i\}\Pr\{A_i\}}{\sum_{j} \Pr\{B|A_j\}\Pr\{A_j\}} \qquad (A.2)$$

In the terminology of Bayesian updating, the probabilities $\Pr\{A_i\}$ are the **prior probabilities** of the events A_i, and $\Pr\{A_i|B\}$ are the **posterior** (or updated) **probabilities** after event B has occurred.

For example, consider the model of searching for food that is known to be located in one of two patches (section 11.1). The two events $A_i =$ "food is in patch i" are mutually exclusive and exhaustive. We assume prior probabilities $p_1 = \Pr\{A_1\} = .55$, and $p_2 = \Pr\{A_2\} = .45$. The detection probabilities are

$p_d(1) = .3$ and $p_d(2) = .6$, where

$$p_d(i) = \Pr\{\text{food is found in patch } i | \text{food is located in patch } i\}$$

Let $B = $ "patch 1 is searched unsuccessfully." Then,

$$\Pr\{B|A_1\} = \Pr\{\text{patch 1 searched unsuccessfully} \mid \text{food is in patch 1}\}$$
$$= 1 - p_d(1)$$
$$\Pr\{B|A_2\} = \Pr\{\text{patch 1 searched unsuccessfully} \mid \text{food is in patch 2}\}$$
$$= 1$$

Hence, the posterior probability is expressed by

$$\Pr\{\text{food is in patch 1} | \text{patch 1 searched unsuccessfully}\}$$
$$= \Pr\{A_1|B\}$$
$$= \frac{\Pr\{B|A_1\}\Pr\{A_1\}}{\Pr\{B|A_1\}\Pr\{A_1\} + \Pr\{B|A_2\}\Pr\{A_2\}}$$
$$= \frac{[1 - p_d(1)]p_1}{[1 - p_d(1)]p_1 + p_2}$$
$$= \frac{.7 \cdot .55}{.7 \cdot .55 + .45}$$
$$= .461$$

The information obtained from a single unsuccessful search of patch 1 reduces the probability that the food is in patch 1 from 55% to 46.1%. Similarly, you can check that a single unsuccessful search of patch 2 would reduce the probability that the food is in patch 2 from 45% to 24.7%. The posterior probability that food is in patch 1, given n_i unsuccessful searches in patch i for $i = 1, 2$ is given in eq. 11.4.

12

Measures of Fitness

In this book, we have usually equated an individual's fitness to its expected lifetime reproductive output. Now we study the concept of fitness in greater detail. The abundant literature on this topic is quite confusing. Therefore, we will start from scratch in an attempt to make the discussion clear and self-contained. Although elementary, the discussion is unfortunately rather lengthy. Briefer discussions of fitness measures occur elsewhere but tend to be incomplete and can be quite confusing.

12.1 Nonoverlapping generations

The basic idea is that fitness is the characteristic of organisms that is maximized by natural selection. But this is too vague—what we really have in mind is an evolutionary contest between two or more alternative traits, or strategies, in a population. Let us begin with the simplest imaginable example, deterministic, density-independent competition between two strategies in a species with nonoverlapping generations. To avoid complications associated with mating and sexual reproduction, we consider only females.

Suppose that the number of i-strategists ($i = 1, 2$) in generation t is $N_i(t)$, and suppose that strategy i results in net fecundity m_i. Then, $N_i(t + 1) = m_i N_i(t)$. Suppose that $m_1 > m_2 > 1$. (Any strategy with $m_i < 1$ will become extinct, and we can ignore this case.) Then, we see that

$$N_i(t) = m_i^t N_i(0) \tag{12.1}$$

where $N_i(0)$ is the population size at $t = 0$. In other words, the i-strategy population grows exponentially, multiplying by the factor m_i in each generation.

Now consider the relative frequency of type i:

$$p_i(t) = \frac{N_i(t)}{N_1(t) + N_2(t)}$$

$$= \frac{m_i^t N_i(0)}{m_1^t N_1(0) + m_2^t N_2(0)} \tag{12.2}$$

248

From the assumption $m_1 > m_2$, it follows that

$$p_1(t) = \frac{N_1(0)}{N_1(0) + (m_2/m_1)^t N_2(0)} \rightarrow \frac{N_1(0)}{N_1(0)} = 1 \quad \text{as } t \rightarrow \infty$$

and similarly $p_2(t) \rightarrow 0$ as $t \rightarrow \infty$. Thus the strategy that has the larger growth rate m_i eventually numerically dominates the alternative strategy, though, in this model, both strategy populations grow infinitely large. (Exactly the same sort of calculation is used in population genetics, except that $p_i(t)$ there represents gene frequency, rather than strategy frequency.)

In this model, fitness can be identified with net reproductive success m_i. The strategy with the larger per capita reproduction m_i wins the evolutionary contest in terms of relative population sizes. The argument extends easily to more than two strategies i.

The assumption underlying eq. 12.1, that each population $N_i(t)$ grows exponentially without bound, is of course not realistic. However, we can immediately extend this argument to include the case in which the entire system is scaled down in each generation by some environmental factor $\Omega_t = \Omega(t, N_1, N_2)$: $N_i(t+1) = \Omega_t m_i N_i(t)$. Iterating this equation gives $N_i(t) = \tilde{\Omega}_t m_i^t N_i(0)$, where $\tilde{\Omega}_t = \Omega_0 \cdot \Omega_1 \cdots \cdots \Omega_{t-1}$. Therefore,

$$p_i(t) = \frac{\tilde{\Omega}_t m_i^t N_i(0)}{\tilde{\Omega}_t m_1^t N_1(0) + \tilde{\Omega}_t m_2^t N_2(0)} = \frac{m_i^t N_i(0)}{m_1^t N_1(0) + m_2^t N_2(0)}$$

which is exactly the same expression (12.2) as before. Hence, fitness can still be identified with fecundity m_i.

In this calculation the assumption is that there is no direct interference or competitive interaction between types—the superior type is simply the one that can dominate numerically. To pass to biologically more realistic situations, we need to consider any or all of the following: mortality risks, density dependence, frequency dependence, iteroparity, environmental stochasticity, sexual reproduction, and genetics. As far as we know, there is as yet no general mathematical theory of evolution that includes all these components, although many publications combine several of them. Here we limit the discussion to the first four items.

Mortality risk

Some parents may die and produce zero offspring, and others survive but produce different numbers of offspring. The mathematics of population growth becomes much more complicated when mortality risk and variable fecundity are considered. In this situation $N_i(t)$ becomes a stochastic process, known in probability theory as a branching process. Branching processes were first considered by F. Galton in 1889, who was interested in the dying out of British aristocratic families. We will not go into the mathematics of branching processes, but a useful starting point is Harris (1963).

Intuitively, we can see what might happen when a rare mutant strategy enters a population that uses a different strategy. Even though, *on average*, the mutant may have greater fecundity, chance extinction may terminate it before it can invade the established strategy. Once the mutant strategy succeeds in establishing itself in some numbers, however, the likelihood of chance extinction becomes negligible (since all members of the mutant population must simultaneously fail to reproduce for the mutant to become extinct). In this case the strategy that has the largest average fecundity will almost certainly win the evolutionary contest.

To express this mathematically, now let m_i denote the average fecundity of an individual (using strategy i) that survives to reproduce, and let σ_i denote the probability that an individual does survive to reproduce. Then the expected, or average, reproduction of a newborn individual is given by

$$\lambda_i = \sigma_i m_i \qquad (12.3)$$

Therefore, this number λ_i is an appropriate measure of fitness under these circumstances and is the basis for adopting expected lifetime reproductive success as a measure of fitness.

To summarize results so far, in a species with nonoverlapping generations and with no density-dependent population regulation, fitness of a strategy can be defined as an individual's expected lifetime reproductive success $\lambda_i = \sigma_i m_i$. This number λ_i is also the average growth rate of a population that uses this strategy. The result also holds under density dependence, provided that all strategies are affected equally by population density. (Obviously λ_i is no longer the growth rate of the population in the latter situation.)

All interesting problems in the study of adaptation are concerned with trade-offs. The above model already displays this feature, if fecundity m_i and survival σ_i are inversely related. For example, by foraging more intensively, a female may be able to increase her fecundity m_i but at the cost of exposing herself to increased predation risk, which decreases survival σ_i. The optimal, fitness-maximizing strategy maximizes the product $\sigma_i m_i$, and thus involves a trade-off between survival and reproduction. Such trade-offs are the core of life-history theory.

Density dependence

Now suppose that the parameters σ_i or m_i (or both) depend on the total population size $N = N(t) = N_1(t) + N_2(t)$. Thus,

$$\lambda_i = \lambda_i(N) = \sigma_i(N)m_i(N) \qquad (12.4)$$

and the equations of population growth become

$$N_i(t+1) = \lambda_i(N)N_i(t) \qquad (12.5)$$

where N denotes the total population of both strategies

$$N = N_1(t) + N_2(t) \tag{12.6}$$

Before continuing, we wish to clarify terminology. In general, the survival and fecundity parameters σ_i and m_i may be (1) density-dependent or independent and (2) frequency-dependent or independent. First, we specify how density and population are related and what these terms mean.

Taking the latter question first, all of the theory presented here deals with population sizes N_i. If the population is uniformly spread over a certain geographical area, then density is always proportional to population size. Otherwise density must have a spatial component—a possibility that we explicitly rule out in our discussion (interesting and important as it may be in ecology). Therefore the term density dependence means that σ_i or m_i depend on population sizes N_1 and N_2. We will use the term **pure density dependence** if σ_i and m_i are functions of the total population size $N = N_1 + N_2$; this is what is assumed in eqs. 12.4 and 12.5.

By contrast, the term **pure frequency dependence** means that $\sigma_i = \sigma_i(f_1)$, etc., where $f_1 = N_1/(N_1 + N_2)$ is the relative frequency of type 1 in the total population. (Since $f_2 = N_2/(N_1+N_2) = 1-f_1$, we don't have to consider both frequencies, at least in the case of two competing types.) In the most general case, $\sigma_i = \sigma_i(N_1, N_2)$, both density and frequency dependence are involved. Henceforth, we will explicitly exclude frequency dependence (which requires a game-theoretic analysis) and assume pure density dependence. There should be no confusion if we just use the term "density dependence" to describe this case.

Now we assume that the functions $\lambda_i(N)$ in eq. 12.4 are decreasing functions of N (see fig. 12.1). The **carrying capacity** K_i for type i is defined by the condition

$$\lambda_i(K_i) = 1 \tag{12.7}$$

If type i is the only strategy present, then $N_i(t)$ will converge to K_i, as $t \to \infty$ (because $N_i(t)$ increases for $N_i(t) < K_i$, and vice versa. We ignore the possibility that K_i may be unstable.) When both strategies are present in the combined population, the strategy that has the larger carrying capacity will persist, and the other strategy is driven to extinction. For example, suppose that the total population size $N(t)$ in fig. 12.1 lies between K_1 and K_2. Then $N_1(t)$ will decrease, and $N_2(t)$ will increase. The total population $N(t) = N_1(t) + N_2(t)$ could either increase or decrease, but eventually $N_2(t)$ will increase to K_2, and $N_1(t)$ will decrease to zero.

To summarize the theory so far, for the case of nonoverlapping generations:

1. If there is no density dependence or if there is density dependence that is not strategy-dependent, then fitness equals $\lambda_i = \sigma_i m_i$, which is equal

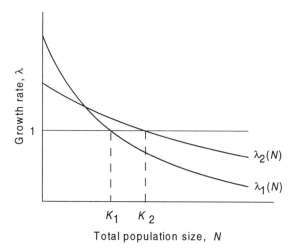

Figure 12.1 Density-dependent growth rates $\lambda_i(N)$ for two competing strategies i. The carrying capacities K_i are determined by the equilibrium condition $\lambda_i(K_i) = 1$. Here, $K_2 > K_1$, and strategy 2 wins the evolutionary contest. Thus, carrying capacity K is a correct fitness measure under density dependence.

to an individual's expected reproductive output. Here, we can ignore the density dependent factor $\Omega(N)$ because it cancels out in the relative frequency ratios p_i of eq. 12.2. Some authors prefer to use $r = \log \lambda$ as the fitness measure in this case.

2. If there is density dependence that is also strategy-dependent, then each λ_i depends on total population size: $\lambda_i = \lambda_i(N)$. In this case, fitness equals the carrying capacity K_i, as defined by the equation $\lambda_i(K_i) = 1$.

This dichotomy was the basis for the concept of r-selection versus K-selection (MacArthur 1962), now largely regarded as oversimplistic (Stearns 1992, p. 206). Nevertheless, the roles of r and K as fitness measures in life-history theory remain important, as will be further explained in the ensuing discussion. Do not confuse r and K, as defined here, with the parameters r and K of the logistic model used in elementary population dynamics, although there is obviously some similarity between the two theories.

To predict the outcome of an evolutionary contest between two (or more) strategies under density plus strategy dependence, therefore it is not sufficient to know single values σ_i and m_i for each strategy. One must also know how these values depend on the total population size N, so that the carrying capacities K_i can be calculated, as in eq. 12.7 .

If one only considers a population at equilibrium $N(t) = \overline{N}$, however, then it is true that fitness $\lambda_i(\overline{N})$ is maximized among potential alternative strategies. For the optimal strategy i, $\lambda_i(\overline{N}) = 1$, and for all other strategies j, $\lambda_j(\overline{N}) < 1$ (see fig. 12.1). In the special case in which the density dependence affects both strategies equally, we have $\lambda_i(N) = \Omega(N) \cdot \lambda_i$ ($i = 1, 2$), where λ_i is

Table 12.1 Density-dependent reproductive strategies

m	$\sigma(m, N)$	$\lambda_m(N) = m\sigma(m, N)$	K_m
1	$\dfrac{0.9}{1 + 0.001N}$	$\dfrac{0.9}{1 + 0.001N}$	0
2	$\dfrac{0.7}{1 + 0.002N}$	$\dfrac{1.4}{1 + 0.002N}$	200
3	$\dfrac{0.6}{1 + 0.006N}$	$\dfrac{1.8}{1 + 0.006N}$	133

independent of N and $\Omega(N)$ is the common density factor. Now the two curves $\lambda_i(N)$ are multiples of one another, and it is clear that $K_1 < K_2$ if and only if $\lambda_1 < \lambda_2$. In this case, therefore, fitness can be defined either as the carrying capacity K or as the density-independent growth rate λ.

An artificial example may help to clarify the situation. Suppose that a female can produce $m = 1, 2,$ or 3 female offspring; these are the three possible strategies. Her survival probability $\sigma(m, N)$ depends on m and also on the total population size N (table 12.1). Having more offspring decreases the parent's chances of survival σ at any population level N, whereas a larger population size N also decreases her survival, especially for larger values of m. In other words, survival is density- and strategy-dependent. The functions $\lambda_m(N)$ are graphed in fig. 12.2.

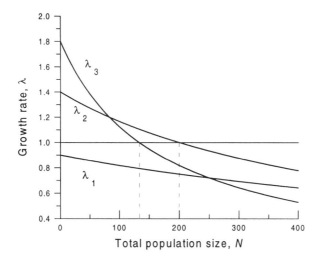

Figure 12.2 Density-dependent growth rates λ_i for the example of table 2.1. The carrying capacities are $K_1 = 0$, $K_2 = 200$, and $K_3 = 133$. Hence, strategy 2 is optimal.

The resulting carrying capacities K_m, obtained by setting $\lambda_m(N) = 1$, are 0, 200, and 133, respectively. (Actually, for $m = 1$, we get $K_1 < 0$, implying that the population becomes extinct; hence in fact, $K_1 = 0$.) Therefore the optimal strategy is $m = 2$. Note that at low density $N \approx 0$, having three offspring is better than having two. Because of the density effects, however, this strategy is ultimately outcompeted by the $m = 2$ strategy.

You may wish to reconsider this example for the case that the density dependence factors are not strategy dependent. For example, let the denominators for each m be $1 + 0.002N$. What is the optimal strategy? What is K?

12.2 Overlapping generations

The calculation of λ_i becomes more complicated for species with overlapping generations. In this case, two different measures of fitness have been proposed. We will describe these two measures and show how they are related. We will also explain when to use which measure. To begin with, we will use the standard notation of life-history theory, but later we will show how to do everything in terms of dynamic programming. The dynamic state variable approach allows us to include the organism's changing state, as well as its behavior, in life-history models.

The basic model of life-history theory was first analyzed by Leslie (1945); the books of Charlesworth (1994); Roff (1992); and Stearns (1992) discuss the theory and application in much detail. To simplify notation, we drop the subscripts i that correspond to different strategies, consider only one strategy, and only females. Fecundity at age x is denoted by m_x, and survival by σ_x (so that σ_x is the probability of surviving from age x to $x + 1$). The numbers m_x and σ_x constitute the **life table** for the population in question.

Let $N_x(t)$ denote the number of individuals of age x in year t. Then

$$N_0(t) = \sum_{x=1}^{\omega} m_x N_x(t) \tag{12.8}$$

$$N_{x+1}(t+1) = \sigma_x N_x(t), \qquad x = 0, 1, \ldots, \omega - 1 \tag{12.9}$$

where ω is the maximum age. These are the basic equations of life-history theory under density independence. It is worth mentioning that the comments made previously regarding branching processes apply here as well. Equation (12.9) tacitly assumes that the population is sufficiently large that all sampling variations in survival can be averaged. We adopt this convention henceforth.

Now we wish to calculate the growth rate λ for this model. To do this, we suppose that the population of newborns $N_0(t)$ grows by some unknown factor λ per generation:

$$N_0(t) = \lambda^t \cdot c \tag{12.10}$$

where the constant c denotes the initial newborn population $N_0(0)$. From eq. 12.9, then

$$N_1(t) = \sigma_0 N_0(t-1) = \sigma_0 \lambda^{t-1} \cdot c$$
$$N_2(t) = \sigma_1 N_1(t-1) = \sigma_1 \sigma_0 \lambda^{t-2} \cdot c = \ell_2 \lambda^{t-2} \cdot c \qquad (12.11)$$
$$\vdots$$
$$N_x(t) = \sigma_{x-1} N_{x-1}(t-1) = \sigma_{x-1} \ldots \sigma_0 \lambda^{t-x} \cdot c = \ell_x \lambda^{t-x} \cdot c$$

where $\ell_x = \sigma_0 \sigma_1 \ldots \sigma_{x-1}$ represents the probability that a newborn survives to age x. The numbers ℓ_x are called the **survival coefficients** or simply the survivorship.

Substituting these expressions in eq. 12.8, we obtain

$$\lambda^t \cdot c = \sum_{x=1}^{\omega} \ell_x m_x \lambda^{t-x} \cdot c$$

or, after simplifying,

$$\sum_{x=1}^{\omega} \ell_x m_x \lambda^{-x} = 1 \qquad (12.12)$$

This is the fundamental equation of life-history theory; it is called the **Euler–Lotka** equation, after the mathematicians who first derived it. The reader may wish to check that the case of nonoverlapping generations, eq. 12.3, is a special case of eq. 12.12, namely, take $\omega = 1$ to obtain $\lambda = \sigma_0 m_1$, which is equivalent to eq. 12.3.

Equation 12.12 has a unique positive solution λ. To explain this, note that the left-hand side is a monotonically decreasing function of λ, whose value approaches $+\infty$ for λ near zero and approaches zero as λ increases to ∞; hence, some unique λ makes the left-hand side equal to one. This value of λ is called the **Malthusian**, or **intrinsic growth rate** of the population. Regardless of the initial age distribution, all age classes $N_x(t)$ grow asymptotically at the same rate λ and reach a **steady** age distribution with (see Gotelli 1995)

$$N_x(t) = \ell_x \lambda^{-x} N_0(t) \qquad (12.13)$$

Given numerical values for σ_x and m_x, the value of the intrinsic growth rate λ determined by eq. 12.12 can be obtained using standard equation-solving computer routines, but we will have no occasion to use them here.

Now consider two or more competing life-history strategies i that have growth rates λ_i. By exactly the same argument as used before for the case of nonoverlapping generations, the strategy that has the largest λ_i will numerically dominate, as $t \to \infty$. Hence, λ_i can be chosen as the fitness measure for strategy i. As before, however, the present model is biologically unrealistic because we have assumed that there is no regulating mechanism to prevent the population from growing exponentially without bounds.

To repeat, in an age-structured population without density dependence, the intrinsic growth rate λ, obtained via the Euler–Lotka equation (12.12), is a valid fitness measure. Many authors use $r = \log \lambda$ instead of λ here; clearly, r is an equally valid fitness measure because $\lambda_1 > \lambda_2$ if and only if $r_1 > r_2$. Also, population viability (or nonextinction) requires $\lambda \geq 1$, or equivalently $r \geq 0$.

To include density dependence in the life-history model, now we introduce the **total weighted population size** $N(t)$:

$$N(t) = \sum_{x=0}^{\omega} w_x N_x(t) \tag{12.14}$$

where $w_x \geq 0$ are given constants. For example, if w_x is the average mass of individuals of age x, then $N(t)$ equals the total biomass of the population. Another example would take $w_x = 0$ for prereproductive ages and $w_x = 1$ for reproductive ages; here, $N(t)$ would represent the total breeding population. This general formulation allows us to consider a variety of possible types of population density effects.

Now we suppose that $\sigma_x = \sigma_x(N)$ and $m_x = m_x(N)$ are density dependent. The Euler–Lotka equation still determines a density-dependent Malthusian growth rate $\lambda(N)$:

$$\sum \ell_x(N) m_x(N) \lambda(N)^{-x} = 1 \tag{12.15}$$

Assuming that survival $\ell_x(N)$ and $m_x(N)$ are decreasing (or at least nonincreasing) functions of N, we conclude that $\lambda(N)$ is also a decreasing function of N. Greater density decreases the growth rate of the population, exactly as in the case of nonoverlapping generations. For two competing strategies, graphs of $\lambda_i(N)$ versus total weighted population size $N = N_1 + N_2$ can be plotted, exactly as in fig. 12.1. Therefore it follows that the fitness of a strategy (i) can once again be defined as its carrying capacity K_i, defined by $\lambda_i(K_i) = 1$. By definition, at the equilibrium, $\overline{N}$ ($\overline{N} = K_2$ in fig. 12.1), individuals are replacing themselves. Thus, at equilibrium, the optimal strategy must satisfy $\lambda_i(\overline{N}) = 1$, and all other strategies require that $\lambda_j(\overline{N}) < 1$.

When $\lambda = \lambda(N) = 1$, the population is at **ecological equilibrium**, and the Euler–Lotka equation (12.12) becomes

$$\sum_{x=0}^{\omega} \ell_x(N) m_x(N) = 1 \tag{12.16}$$

The term $\ell_x(N) m_x(N)$, the product of the probability of surviving to age x and fecundity at age x (in a population of size N), equals the expected reproduction at future age x of a newly born female. The left-hand side of this equation equals the total expected lifetime reproduction of a newborn female. This is usually denoted by R_0:

$$R_0 = R_0(N) = \sum_{x=0}^{\omega} \ell_x(N)m_x(N) \tag{12.17}$$

In a population at equilibrium, each newborn female produces, on average, one surviving female offspring over her life span, so that $R_0(\overline{N}) = 1$.

Imagine two or more life-history strategies i, that have intrinsic growth rates $\lambda_i(N)$ and expected lifetime reproductions $R_{0i}(N)$. Graphs of $R_{0i}(N)$ can be drawn, analogous to the graphs of $\lambda_i(N)$ in fig. 12.1. Now the various carrying capacities K_i are given by $R_{0i}(K_i) = 1$, and the optimal strategy is the one that has the largest carrying capacity K_i. Moreover, at this double equilibrium K^* (population size and life-history strategy both in equilibrium), we see that $R_{0i}(K^*)$ is maximized:

$$R_{0j}(K^*) < R_{0i}(K^*) \quad \text{for all suboptimal strategies } j$$

This is exactly analogous to the case for $\lambda_i(K^*)$ as already explained. Our results can be summarized as follows:

Under pure density dependence, the optimal life-history strategy maximizes its carrying capacity K. When the (weighted) population (N) is in equilibrium at this maximal carrying capacity, both $\lambda(N)$ and $R_0(N)$ are maximized (and equal to 1) with respect to strategy.

We can expand the numerical example used earlier in table 12.1, to illustrate this conclusion. Suppose, for simplicity, that fecundity m and survival $\sigma(m)$ are not age-dependent: $m_x = m$ and $\sigma_x = \sigma(m)$. Then $\ell_x = \ell_x(m) = (\sigma(m))^x$ and the Euler–Lotka equation (12.12) becomes

$$\sum_{x=1}^{\infty} \ell_x m_x \lambda^{-x} = m \sum_{x=1}^{\infty} \sigma(m)^x \lambda^{-x} = m \sum_{x=1}^{\infty} \left[\frac{\lambda}{\sigma(m)} \right]^{-x}$$

$$= \frac{m}{\lambda/\sigma(m) - 1} = 1$$

Hence,

$$\lambda = (m+1)\sigma(m) \tag{12.18}$$

that is, in each generation each female produces, on average, $m\sigma(m)$ offspring and survives herself with probability $\sigma(m)$. Also,

$$R_0 = \sum_{x=1}^{\infty} \ell_x m_x = m \sum_{1}^{\infty} [\sigma(m)]^x = \frac{m\sigma(m)}{1 - \sigma(m)} \tag{12.19}$$

When survival is density-dependent and $\sigma(m) = \sigma(m, N)$ as given in table 12.1, the optimal strategy can be obtained either from the conditions that

Table 12.2 Calculation of carrying capacity K, using both λ and R_0 fitness measures

m	$\sigma(m, N)$	$\lambda_m(N)$	$K_\lambda(m)$	$R_0(m, n)$	$K_{R_0}(m)$
1	$\dfrac{0.9}{1 + 0.001N}$	$\dfrac{1.8}{1 + 0.001N}$	800	$\dfrac{0.9}{0.1 + 0.001N}$	800
2	$\dfrac{0.7}{1 + 0.002N}$	$\dfrac{2.1}{1 + 0.002N}$	550	$\dfrac{1.4}{0.3 + 0.002N}$	550
3	$\dfrac{0.6}{1 + 0.006N}$	$\dfrac{2.4}{1 + 0.006N}$	233	$\dfrac{1.8}{0.4 + 0.006N}$	233

$\lambda(m, K) = 1$ and K is maximized or that $R_0(m, K) = 1$ and K is maximized. In fact, the equations $\lambda(m, K) = 1$ and $R_0(m, K) = 1$ necessarily have the same solution K, which depends on the strategy m. The calculations using eqs. 12.18 and 12.19 are given in table 12.2. The optimal strategy by either criterion is $m = 1$. This can be compared with the optimal strategy at low density ($N \approx 0$), which is $m = 3$, and with the optimal strategy in a density-dependent non-age-structured model, which is $m = 2$. It may be a worthwhile review exercise to try to explain to yourself the biological interpretation of these three cases.

The foregoing discussion and example show that, in fact, $r = \log(\lambda)$ and R_0 are both correct fitness measures under density dependence, but *they must both be evaluated at the optimal equilibrium $N = K$*. When using either fitness measure in a model, it must be realized that *the density dependence of every strategy must be explicitly considered* (see Mylius and Diekmann 1995).

Since unlimited population growth is biologically unrealistic, the use of density-independent models is of dubious value. There is one special situation, however, in which r (and not R_0) is the correct fitness measure—namely, suppose that the density-dependent factors are both age and strategy independent, that is, these factors are the same for all age classes and all strategies. As in the case of nonoverlapping generations discussed earlier, then r (or λ) assessed at low population density ($N \approx 0$) is a correct fitness measure, and R_0 ($N \approx 0$) is not correct. This remains true even if the density-dependent factors are stochastic in time. Interesting as this special case may be, we emphasize that it holds only under the assumptions of complete age and strategy independence of the density factors.

This completes our discussion of fitness measures in life-history theory. Note that we have not considered at all the problems of frequency dependence and stochastic environments. The theory becomes much more esoteric in these situations. A full theory of fitness measures would include the following

1. age structure
2. density dependence

3. strategy dependence
4. frequency dependence
5. stochastic environments.

To our knowledge, no one has produced a usable theory encompassing all of these dimensions. Some partial theories dealing with various points are

1,2,3: this chapter

2,3,4,5: Kisdi and Meszena (1993)

1,3,5: Orzack (1993) and a long series of papers by Tuljapurkar and Orzack, some cited in Orzack (1993).

Metz et al. (1992) considered the most general situation, but their criterion, the "dominant Lyapounov exponent," is probably not operational, except for the known special cases.

12.3 Fitness measures in dynamic state variable models

In this book, we have usually defined the fitness function $F(x, t)$ as total future lifetime expected reproduction from period t on, given that $X(t) = x$. This corresponds to the life-history fitness measure R_0.

To be specific, suppose we are developing a dynamic state variable model for an iteroparous species, using field data for estimating the parameters. Let x denote the age of an individual $(x = 0, 1, \ldots, w)$, and let $Y(x)$ be its state at the beginning of age x. The fitness function is defined as

$$F(y, x) = \text{maximum expected total reproduction from}$$
$$\text{age } x \text{ to } w \text{ inclusive, given that } Y(x) = y$$

(Note that x here plays the same role as time t in the notation used elsewhere in this book.) If $Y(0) = y_0$ is the state of a newborn female, then $F(y_0, 0) = R_0$ since both values represent the total expected lifetime reproductive success of a newborn female. For a stable population therefore, we require that

$$F(y_0, 0) = 1 \tag{12.20}$$

This important condition provides a consistency check on the model. If the computed value of $F(y_0, 0)$ turns out to be different from one, then the model predicts a growing or declining population, which is inconsistent with the use of R_0 as the fitness measure. Something needs to be changed—either the model parameters are incorrect, or we should use r as the fitness measure. First, we consider a numerical example to illustrate how a model might reasonably be "tuned" to ensure that $R_0 = F(y_0, 0) = 1$. The possibility of using r as fitness in a dynamic state variable model is discussed later.

Imagine that the survival coefficients $\sigma(1) = 0.9$, $\sigma(2) = 0.7$, and $\sigma(3) = 0.6$ for clutch sizes $m = 1, 2, 3$ have been measured for a certain population, which is thought to be approximately at equilibrium. From eq. 12.19, the R_0 values are $R_0 = 9$, 4.67, and 4.5 for $m = 1, 2, 3$, respectively. Thus, the optimal reproductive strategy is $m = 1$, giving $R_0 = 9$. This can't be correct in a stable population.

If these were real data, the conclusion would be either that the population is far from equilibrium (or at least that the subpopulation from which the estimates of $\sigma(i)$ were taken is not at equilibrium) or else that we have neglected something in the model. Upon reconsidering the model and the data, we realize in fact that we have neglected juvenile mortality; that is, a successful clutch of $m = 3$ females does not, on average, result in adding three females to the breeding population next year. Rather than the survival function $\ell_x(m) = [\sigma(m)]^x$ used in obtaining eq. 12.19, we should have used $\ell_x(m) = \sigma_0[\sigma(m)]^{x-1}$, where σ_0 is juvenile mortality, here assumed independent of m. Then, eq. 12.19 becomes

$$R_0 = \sum_1^\infty \ell_x m_x = \frac{m\sigma_0}{1 - \sigma(m)}$$

This is equal to $10\sigma_0$, $6.67\sigma_0$, and $7.5\sigma_0$ for $m = 1, 2, 3$. Again, the optimal reproductive strategy is $m = 1$; for a stable population we require that $R_0 = 1$, so that $\sigma_0 = 0.1$. With any other value of σ_0, our choice of fitness measure R_0 would not be logically consistent.

In this example, tuning σ_0 to obtain $R_0 = 1$ does not alter the prediction of the model. This may not be true in other models, as the next example shows.

A model of optimal foraging effort

Imagine a forager that reproduces once at each age $x = 0, 1, \ldots \omega$. The forager can spend a proportion h of her time foraging ($0 \leq h \leq 1$); h is called foraging effort and is the decision variable. Reproduction m and mortality risk μ are increasing functions of foraging effort:

$$m = m(h), \text{ where } m(0) = 0, m' > 0$$

$$\mu = \mu(h), \text{ where } \mu(0) \geq 0, \mu' > 0$$

To determine the age-dependent optimal foraging strategy corresponding to the fitness measure R_0, we introduce the fitness function

$$F(x) = \text{maximum expected reproduction from} \qquad (12.21)$$
$$\text{age } x \text{ to } \omega \text{ inclusive}$$

In this example there is no state variable $Y(x)$ at all. The dynamic programming algorithm is

$$F(\omega + 1) = 0 \qquad (12.22)$$

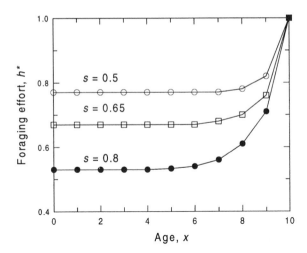

Figure 12.3 Optimal foraging effort h^* as a function of age x, using R_0 maximization, for three values of the maximum survival rate s. Only the case $s = 0.65$ (which implies that $R_0 = 1.0$) is logically consistent.

and, for $0 \le x \le \omega$

$$F(x) = \max_h \sigma(h)[(m(h) + F(x+1)]$$ (12.23)

where $\sigma(h) = 1 - \mu(h)$. For our example, we take $\sigma(h) = se^{-kh}$ and $m(h) = bh$, where s denotes the maximum possible survival probability per period, k is a parameter determining the effect of effort h on survival, and b is the maximum possible reproduction per period.

Figure 12.3 shows the optimal effort levels $h^*(x)$ for this example, with parameter values $k = 1$, $b = 3$, and $\omega = 10$ and for three values of s. The values of $R_0 = F(0)$ are

s	R_0
0.80	1.42
0.65	1.00
0.50	0.70

Thus, assuming that the values of k, b, and ω are "correct," the value of s should be tuned to $s = 0.65$ for consistency of the model. Figure 12.3 shows that the quantitative predictions of the model depend on the values of s, so the question of consistency is important.

The lesson from this example is that in any life-history or dynamic state variable model that uses R_0 maximization, the consistency condition $R_0 = 1$ must be met, if necessary by tuning some of the model parameters. Depending on which parameters are tuned, the model predictions may depend on this tuning.

(Although this discussion and example are highly simplified for ease of understanding, much more complicated whole-life models can be constructed, involving bona fide state variables, basic time periods t shorter than one generation, and so on. The consistency condition $R_0 = 1$ is still required when fitness is defined in terms of total expected lifetime reproduction. An example was given in chapter 6.)

For further discussion of state-dependent life-history theory see McNamara and Houston (1996).

Maximization of r

Finally, we consider the problem of maximizing r in a dynamic state variable model. McNamara (1992) has shown that this can be accomplished by dynamic programming if one defines a new fitness function by

$$F(y, x, t) = \begin{aligned} &\text{maximum expected total number of descendants} \\ &\text{(including self) alive at a given future time } T, \\ &\text{for an individual with state } y \text{ and age } x \text{ at time } t \end{aligned} \quad (12.24)$$

Then, for $t \ll T$, the stationary strategy that optimizes $F(y, x, t)$ also maximizes the population growth rate r. This result, though not trivial mathematically, is intuitively appealing—the strategy that leads to the largest number of descendants in the distant future maximizes r. Descendants are defined as female offspring of all ages, plus the parent herself if she is still alive.

As with the conventional R_0-based fitness function, it is possible to derive a dynamic programming equation for $F(y, x, t)$ and to solve it by computer. Now we do so for the previous model of optimal foraging effort. Since there is no state variable y in this model, the r-based fitness function simplifies to

$$F(x, t) = \begin{aligned} &\text{maximum expected total number of descendants} \\ &\text{produced between } t \text{ and } T \text{ and alive at the end of} \\ &\text{period } T, \text{ for an individual of age } x \text{ at time } t \end{aligned} \quad (12.25)$$

Since ω is the last age of reproduction,

$$F(\omega + 1, t) = 0 \quad (12.26)$$

for any t. For ages $x < \omega$, at $t = T$,

$$F(x, T) = \max_h \sigma(h) m(h) \quad (12.27)$$

because there is no further accumulation after T. For times previous to T, with reproductive effort h, the female produces $m(h)$ offspring, each of whom produces $F(0, t+1)$ descendants at time T (by definition, eq. 12.25), and she herself survives to time $t + 1$ with probability $\sigma(h)$. Hence,

$$F(x, t) = \max_h \sigma(h)[m(h)F(0, t+1) + F(x+1, t+1)] \quad (12.28)$$

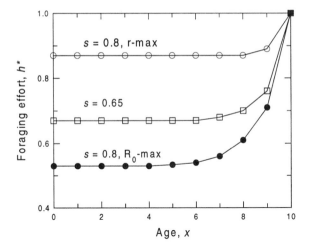

Figure 12.4 Optimal foraging effort h^* as a function of age x, using r maximization and R_0 maximization, with $s = 0.8$ and $s = 0.65$. When $s = 0.65$, the population is stable, and the two calculations yield identical results. Otherwise, the population is unstable, and different results are obtained.

Figure 12.4 shows the optimal effort levels $h^*(x)$ for both cases, r maximization and R_0 maximization. When $\sigma_0 = 0.65$, we know from the R_0 maximization results above that $R_0 = 1$, that is, the population is stable. Hence, also $\lambda = 1$ (or $r = 0$). As pointed out earlier, R_0 and r maximization are equivalent in this case. The middle curve in fig. 12.4 shows $h^*(x)$ computed by either algorithm.

The uppermost curve in fig. 12.4 gives $h^*(x)$ for $\sigma_0 = 0.8$, using r maximization, and the lowest curve gives $h^*(x)$ using R_0 maximization (which is of course incorrect for this case). The computed value of $r = \log \lambda$ is $r = 0.94$; the population in this case would increase by a factor of $e^r - 1 = 2.56$ in each generation (unless controlled by factors not age- or strategy-dependent). It is typical for a growing population ($r > 0$) that reproductive effort predicted by r maximization is higher than that predicted by R_0 maximization because of the advantage gained by earlier reproduction.

This example illustrates the two major points of this section: first, that r maximizing life-history or behavioral strategies (including state dependent strategies) can readily be computed using dynamic programming; and second, that R_0- and r maximizing strategies are identical for stable populations but differ for nonstable populations (where r is the correct fitness measure).

Offspring quality

The r maximization algorithm is required (even for stable populations) when parental decisions affect offspring quality. For example, the clutch size laid on a host by a parasitic wasp affects the average size of her offspring. Larger offspring have greater reproductive success than smaller offspring (see chapter 4).

In such cases, merely counting the total number of offspring clearly gives an incorrect measure of fitness. Newborn offspring themselves have states, say z, which are under the control of their parent's behavior. Therefore, the parent's fitness is defined in terms of the total number of descendants at some remote future time T. In a stable population, this number will equal one for the optimal strategy; alternative strategies will yield fitness less than one.

Parental care

Although we include no such models in this book, parental care can also be modeled using the dynamic state variable method. Since parental care affects offspring fitness, once again the correct fitness function is given in terms of the total number of descendants at future time T.

Stochastic environments

An important assumption in this theory is that the environment does not fluctuate from year to year. However, as we have noted, the use of r maximization is correct if the environment does fluctuate, provided that the environmental effects are age- and strategy-independent.

Life-history strategies that are adaptive to stochastic environments have been studied in an extensive literature (reviewed in Yoshimura and Clark 1993). Recently McNamara (1997; also McNamara et al. 1995) has shown how to solve dynamic state variable models in stochastic environments. An important feature is that individual optimization models are no longer appropriate. We refer the interested reader to the book of Houston and McNamara (1999). Grafen (1999) discusses McNamara's result in terms of fitness maximizing individuals and total reproductive value.

Appendix

Programs available on the OUP Web site

The following programs can be downloaded from the Oxford University Press Web site: http://www.oup-usa.org/sc/0195122674/ or /sc/0195122666/.

True Basic^{TM} Programs

At the Web site, you will find the following models programmed in True Basic^{TM}:

Patch-Selection Model (chapter 1)

Body-Size Model (section 2.7)

Proovigenic Parasitoid Model (section 4.4)

BASIC was created by John Kemeny (a famous applied mathematician) and Thomas Kurtz at Dartmouth College in 1964. About 20 years later, they pretty much asked, "Computing has changed so much in the last 20 years, if we were now creating BASIC, what would we do?" The result was True Basic^{TM}, which has the following features: it contains the full range of modern programming structures; it allows one to use subroutines and externally defined functions; and it allows one to do recursions, contains terrific graphics packages (indeed, that is one reason why MM sticks with it), and has built-in matrix operations.

True Basic^{TM} comes in PC, Mac, and Unix versions and is very easy to use. Furthermore, it is very easy to learn. If you are a neophyte programmer, we encourage you to purchase the Student Edition of True Basic^{TM}; with effort you can be programming in one afternoon. Because of its advanced structural features, this language is also very good for advanced programmers (MM has used it since 1985, even though he knows FORTRAN, C, and C^{++}).

C^{++} Programs

The C^{++} programs available at the OUP Web site are

Patchnew.cpp. A C^{++} version of the code for the basic patch-selection model of chapter 1. It also performs forward iteration as described in section 2.3.

Wtit1.cpp. The willow tit foraging model of section 5.1.

PatchESS.cpp. Patch model with frequency dependence—section 10.2.

Life-his.cpp. Program to compare r- and R_0 maximization—chapter 12.

You will need a C^{++} compiler to run these programs: I (CWC) used Borland Turbo C^{++}. The programs are not object oriented, and use only a few C^{++} features that are improvements over the C language.

Array indexing

The indexing of arrays in the C^{++} programs may be confusing. An array A in C^{++} is always indexed starting at zero, so that the values A[0], A[1],...,A(I) are stored in the computer. These values can be accessed in two different ways: A[i]= *(A+i). The expression *(A+i) instructs the computer to access the position in memory that is i data units beyond the beginning position of the array A. The difficulty arises in two-dimensional (or higher dimensional) arrays. C^{++} does not recognize notation such as A[i,j]—or at least my C^{++} compiler doesn't. Instead, you have to specify how the computer orders the entries A[i,j] into a single list. All of the programs do this in the same way.

Specifically, for an array A[i,j] with i = 0,1,...,I and j= 0,1,..., J, the programs express A[i,j] as A[i,j] = *(A + addr(i,j)), where addr(i,j) is coded separately, and is given by addr $(i, j) = i * (J + 1) + j$. This means that in the computer, the values A[i,j] are stored in the order A[0,0], A[0,1],...,A[0,J], A[1,0], A[1,1],...,A[1,J],...,A[I,0], A[I,1],...,A[I,J].

References

Aarts, E., and Korst, J. (1989) *Simulated annealing and Boltzmann machines.* John Wiley & Sons, New York.

Abrahams, M.V., and Dill, L.M. (1989) A determination of the energetic equivalence of the risk of predation. *Ecology* **70**, 999–1007.

Adler, F., and Nuernberger, B. (1994) Persistence in patchy irregular landscapes. *Theoretical Population Biology* **45**, 41–75.

Ahlquist, J.E., Bledsoe, A.H., Ratti, J.T., and Sibley, C.G. (1987) Divergence of the single-copy DNA sequences of the western grebe (*A. occidentalis*) and the Clark's grebe (*A. clarkii*) as indicated by DNA-DNA hybridization. *Postilla* No. 200.

Alerstam, T. (1979) Optimal use of wind by migrating birds: combined drift and overcompensation. *Journal of Theoretical Biology* **79**, 341–353.

Alerstam, T. (1990) *Bird migration.* Cambridge University Press, Cambridge.

Alerstam, T. (1991) Bird flight and optimal migration. *Trends in Ecology & Evolution* **6**, 210–215.

Alerstam, T., and Hedenström, A. 1998. Optimal migration. *Journal of Avian Biology* **29**, 337–636.

Alerstam, T., and Lindström, Å. (1990) Optimal bird migration: the relative importance of time, energy and safety. In E. Gwinner (ed.), *Bird migration: Physiology and ecophysiology.* Springer-Verlag, Berlin, pp. 331–351.

Alverson, D.L., Freeberg, M.H., Murawski, S.A. and Pope, J.G. (1994) *A global assessment of fisheries bycatch and discards.* FAO Fisheries Technical Paper 339, Food and Agriculture Organization of the United Nations, Rome.

Anderies, J.M. (1996) An adaptive model for predicting !kung reproductive performance: A stochastic dynamic programming approach. *Ethology and Sociobiology* **17**, 221-245.

Andersson, M., and Krebs, J.R. (1978) On the evolution of hoarding behavior. *Animal Behavior* **26**, 707–711.

Bai, B.R., Luck, R.F., Forster, L., Stephens, B. and Janssen, J.A.M. (1992) The effect of host size on quality attributes of the egg parasitoid *Trichogramma pretiosum. Entomologia Experimentalis et Applicata* **64**, 37–48.

Baker, J., and Whelan, R.J. (1994) Ground parrots and fire at Barren Grounds, New South Wales—a long-term study and an assessment of management implications. *Emu* **94**, 300–304.

Balda, R.P., and Kamil, A.C. (1989) A comparative study of cache recovery by three corvid species. *Animal Behavior* **38**, 486–495.

267

Baldi, P., and Brunak, S. (1998) *Bioinformatics: The machine learning approach.* MIT Press, Cambridge, Mass.

Barta, Z., Flynn, R., and Giraldeau, L.-A. (1997) Geometry for a selfish foraging group: a genetic algorithm approach. *Proceedings of the Royal Society of London* B **264**, 1233–1238.

Bartlett, B.R. (1964) Patterns in the host-feeding habit of adult parasitic hymenoptera. *Annals of the Entomological Society of America* **57**, 344–350.

Beasley, D., Bull, D.R., and Martin, R.R. (1993a) An overview of genetic algorithms. Part I, fundamentals. *University Computing* **15**, 58–69.

Beasley, D., Bull, D.R., and Martin, R.R. (1993b) An overview of genetic algorithms: Part 2, research topics, *University Computing* **15**, 170–181.

Bednekoff, P.A. (1996) Translating mass dependent flight performance into predation risk: an extension of Metcalfe & Ure. *Proceedings of the Royal Society of London* B **263**, 887–889.

Bednekoff, P.A. (1997) Mutualism among safe, selfish sentinels: A dynamic game. *American Naturalist* **150**, 373–392.

Bednekoff, P.A., and Balda, R.P. (1996) Social caching and observational spatial memory in pinyon jays. *Behaviour* **133**, 807–826.

Bednekoff, P.A., and Houston, A.I. (1994a) Avian daily foraging patterns: effects of digestive constraints and variability. *Evolutionary Ecology* **8**, 36–52.

Bednekoff, P.A., and Houston, A.I. (1994b) Optimizing fat reserves over the entire winter: a dynamic model. *Oikos* **71**, 408–415.

Begon, M., Harper, J.L., and Townsend, C.R. (1990) *Ecology: Individuals, populations and communities.* Blackwell Scientific Publications, Boston, Mass.

Berger, J.O., and Berry, D.A. (1988) Statistical analysis and the illusion of objectivity. *American Scientist* **76**, 159–165.

Berkson, J.M. (1996) Modeling the restoration of a metapopulation: Implications for resource management. Ph.D. Dissertation. Montana State University, Bozeman, Mont.

Bernstein, P. (1996) *Against the gods. The remarkable story of risk.* John Wiley & Sons, New York.

Bishop, C.M. (1995) *Neural networks for pattern recognition.* Oxford University Press, New York.

Blurton Jones, N. (1986) Bushman birth spacing: a test for optimal interbirth intervals. *Ethology and Sociobiology* **7**, 91–105.

Blurton Jones, N. (1987) Bushman birth spacing: direct tests of some simple predictions. *Ethology and Sociobiology* **8**, 183–203.

Blurton Jones, N.G., and Sibly, R.M. (1978) Testing adaptiveness of culturally determined behaviour: Do Bushman women maximize their reproductive success by spacing births widely and foraging seldom? Society for the Study of Human Biology Symposium 18: *Human Behaviour and Adaptation.* Taylor and Francis, London, pp. 135–157.

Borgerhoff Mulder, M., and Judge, D.S. (1993) Sex, statistical reasoning and other human interests. *Trends in Ecology and Evolution* **8**, 6–7.

Bouskila, A., and Blumstein, D.T. (1992) Rules of thumb for predation hazard assessment: predictions from a dynamic model. *American Naturalist* **139**, 161–176.

Brodin, A. (1992) Cache dispersion affects retrieval time in hoarding Willow Tits. *Ornis Scandinavica* **23**, 7–12.

Brodin, A. (1993) Radio-ptilochronology—tracing radioactively labelled food in feathers. *Ornis Scandinavica* **24**, 167–173.

Brodin, A. (1994) The role of naturally stored food supplies in the winter diet of the boreal Willow Tit *Parus montanus*. *Ornis Svecica* **4**, 31–40.

Brodin, A., and Clark, C.W. (1997) Long-term hoarding in the *Paridae*: A dynamic model. *Behavioral Ecology* **8**, 178–185.

Brodin, A., and Ekman, J. (1994) Benefits of food hoarding. *Nature* **372**, 510.

Bruderer, B., and Weitnauer, E. (1972) Radarbeobachtungen über Zug und Jachtflüge des Mauerseglers (*Apus apus*). *Rev. Suisse Zool.* **79**, 1190–1200.

Bull, C.D., Metcalfe, N.B., and Mangel, M. (1996) Seasonal matching of foraging to anticipated energy requirements in anorexic juvenile salmon. *Proceedings of the Royal Society of London* B **263**, 13–18.

Bullock, J.M., Hill, B.C., and Silvertown, J. (1994) Demography of *Cirsium vulgare* in a grazing experiment. *Journal of Ecology* **82**, 101–111.

Butler, R.W., Williams,T.D., Warnock, N., and Bishop, M.A. (1997) Wind assistance: a requirement for migration of shorebirds? *Auk* **114**, 456–466.

Byrd, J.W., Houston, A.I., and Sozou, P.D. (1991) Ydenberg's model of fledging time—a comment. *Ecology* **72**, 1893–1896.

Casas, J. (1988) Analysis of searching movements of leafminer parasitoids in a structured environment. *Physiological Entomology* **13**, 373–380.

Casas, J. (1989) Foraging behaviour of a leafminer parasitoid in the field. *Ecological Entomology* **14**, 257–265.

Casas, J., Gurney, W.S.C., Nisbet, R., and Roux, O. (1993) A probabilistic model for the functional response of a parasitoid at the behavioural time-scale. *Journal of Animal Ecology* **62**, 194–204.

Chan, M.S. (1991) Host feeding in parasitic wasps: A study of population patterns generated by individual behaviour. Ph.D. Thesis, Department of Biology, Imperial College, U.K.

Charlesworth, B. (1994) *Evolution in age-structured populations*, 2nd edition. Cambridge University Press, Cambridge, U.K.

Charnov, E.L. (1976) Optimal foraging: the marginal value theorem. *Theoretical Population Biology* **9**, 129-136.

Chesson, P., and Murdoch, W.W. (1986) Aggregation of risk: relationships among host-parasitoid models. *American Naturalist* **127**, 696–715.

Clark, C.W. (1990) *Mathematical bioeconomics*. 2nd edition. Wiley Interscience, New York.

Clark, C.W. (1993) Dynamic models of behavior: An extension of life history theory. *Trends in Ecology and Evolution* **8**, 205–209.

Clark, C.W., and Butler, R.W. (1999) Fitness components of avian migration: A dynamic model of Western Sandpiper migration. *Evolutionary Ecology Research* **1**, 443–457.

Clark, C.W., and Ekman, J. (1995) Dominant and subordinate fattening strategies: a dynamic game. *Oikos* **72**, 205–212.

Clark, C.W., and Harvell, C.D. (1992). Inducible defenses and the allocation of resources: a minimal model. *American Naturalist* **139**, 521–539.

Clark, C.W., and Levy, D.A. (1988) Diel vertical migrations by juvenile sockeye salmon and the antipredation window. *American Naturalist* **131**, 271–290.

Clark, C.W., and Mangel, M. (1984) Foraging and flocking strategies: information in an uncertain environment. *American Naturalist* **123**, 626–641.

Clark, C.W., and Mangel, M. (1986) The evolutionary advantages of group foraging. *Theoretical Population Biology* **30**, 45–75.

Clark, C.W., and Rosenzweig, M.L. (1994) Extinction and colonization processes: parameter estimates from sporadic surveys. *American Naturalist* **143**, 583–596.

Clark, C. W., and Ydenberg, R.C. (1990a) The risks of parenthood. I. General theory and applications. *Evolutionary Ecology* **4**, 21–34.

Clark, C.W., and Ydenberg, R.C. (1990b) The risks of parenthood. II. Parent-offspring conflict. *Evolutionary Ecology* **4**, 312–325.

Collier, T.R. (1995a) Adding physiological realism to dynamic state variable models of parasitoid host feeding. *Evolutionary Ecology* **9**, 217–235.

Collier, T.R. (1995b) Host feeding, egg maturation, resorption, and longevity in the parasitoid *Aphytis melinus* (Hymenoptera: Aphelinidae). *Annals of the Entomological Society of America* **88**, 206–214.

Collier, T.R., Murdoch, W.W., and Nisbet, R.M. (1994) Egg load and the decision to host-feed in the parasitoid, *Aphytis melinus*. *Journal of Animal Ecology* **63**, 299–306.

Corbett, A., and Plant, R.E. (1993) Role of movement in the response of natural enemies to agroecosystem diversification—a theoretical evaluation. *Environmental Entomology* **22**, 519–531.

Costantino, R.F., and Desharnais, R.A. (1991) *Population dynamics and the Tribolium model: Genetics and demography.* Springer-Verlag, New York.

Crawford, C.B. (1993) The future of sociobiology—counting babies or studying proximate mechanisms? *Trends in Ecology and Evolution* **8**, 183-186.

Crowley, P.H., and Hopper, K.R. (1994) How to behave around cannibals: a density-dependent dynamic game. *American Naturalist* **143**, 117–154.

Cunjak, R.A., and Power, G. (1987) The feeding and energetics of stream-resident trout in winter. *Journal of Fish Biology* **31**, 493–511.

Davis, L. (1987) *Genetic algorithms and simulated annealing.* Pitman, London, U.K.

Davis, L. (1991) *Handbook of genetic algorithms.* Van Nostrand Reinhold, New York.

Dawkins, R., and Carlisle, T.R. (1976) Parental investment, mate desertion and a fallacy. *Nature* (London) **262**, 131–133.

de Bach, P. (1943) The importance of host-feeding by adult parasites in the reduction of host populations. *Journal of Economic Entomology* **36**, 647–658.

de Jong, T., Klinkhamer, P.G.L., Geritz, S.A.H., and van der Meijden, E. (1989) Why biennials delay flowering: an optimization model and field data on *Cirsium vulgare* and *Cynoglossum officinale*. *Acta Botanica Nederland* **38**, 41–55.

Dixit, A.K., and Pindyck, R.S. (1994) *Investment under uncertainty.* Princeton University Press, Princeton, N.J.

Dukas, R. (1998) Evolutionary ecology of learning. In R. Dukas (ed.), *Cognitive Ecology*. University of Chicago Press, Chicago, pp. 129–174.

Efron, B., and Tibshirani, R.J. (1993) *An introduction to the bootstrap.* Chapman and Hall, New York.

Eichhorst, B.A. (1994) Mitochondrial DNA variation within the Western-Clark's Grebe complex (*Podicipedidae: Aechmophorus*). Ph.D. Dissertation, University of North Dakota, Grand Forks, N. Dak.

Ekman, J. (1979) Coherence, composition and territories of winter social groups of the Willow Tit *Parus montanus* and the Crested Tit *P. cristatus*. *Ornis Scandinavica* **10**, 56–58.

Ekman, J. (1990) Alliances in winter flocks of willow tits; effects of rank on survival and reproduction success in male-female associations. *Behavioral Ecology and Sociobiology* **26**, 239–245.

Ekman, J., and Lilliendahl, K. (1993) Using priority to food access: fattening strategies in dominance-structured willow tit (*Parus montanus*) flocks. *Behavioral Ecology* **4**, 232–238.

Emlen, J.M. (1966) The role of time and energy in food preference. *American Naturalist* **100**, 611–617.

Emlen, S.T. (1995) An evolutionary theory of the family. *Proceedings of the National Academy of Sciences USA* **92**, 8092–8099.

Endler, J. (1986) *Natural selection in the wild*. Princeton University Press, Princeton, N.J.

Ens, B.J., Piersma, T., and Tinbergen, J.M. (1994) Towards predictive models of bird migration schedules: Theoretical and empirical bottlenecks. NIOZ-Rapport 1994–5, Netherlands Institute for Sea Reseach, Den Burg, The Netherlands.

Falconer, D.S. (1993) *Introduction to quantitative genetics*. Longman Scientific and Technical, London, U.K.

Farmer, A.H., and Wiens, J.A. (1998) Optimal migration schedules depend on the landscape and the physical environment: a dynamic modeling view. *Journal of Avian Biology* **29**, 405–415.

Farmer, A.H., and Wiens, J.A. (1999) Models and reality: a test of time and energy tradeoffs in Spring migration of the pectoral sandpiper (*Calidris melanotos*). *Ecology* (in press).

Feynman, R.P. (1965) *The Feynman lectures on physics*. Freeman, San Francisco.

Fletcher, J.P., Hughes, J.P., and Harvey, I.F. (1994) Life expectancy and egg load affect oviposition decisions of a solitary parasitoid. *Proceedings of the Royal Society of London* B **258**, 163–167.

Forbes, L.S., and Ydenberg, R.C. (1992) Sibling rivalry in a variable environment. *Theoretical Population Biology* **41**, 135–160.

Fretwell, S.D., and Lucas, H.L. (1970) On territorial behaviour and other factors influencing habitat distribution in birds. I. Theoretical development. *Acta Biotheoretica* **19**, 16–36.

Friend, G.R. (1993) Impact of fire on small vertebrates in mallee woodlands and heathlands of temperate Australia: a review. *Biological Conservation* **65**, 99–114.

Gardiner, W.R., and Geddes, P. (1980) The influence of body composition on the survival of juvenile salmon. *Hydrobiologia* **69**, 67–72.

Gaston, A.J. (1992) *The ancient murrelet*. Poyser Press, London.

Gaston, A.J. (1998) Modelling departure strategies in auks. *Auk* **115**, 798–800.

Gill, A.B., and Hart, P.J.B. (1996) How feeding performance and energy intake change with a small increase in the body size of the three-spined stickleback. *Journal of Fish Biology* **48**, 878–890.

Gillis, D.M., Pikitch, E.K., and Peterman, R.M. (1995a) Dynamic discarding decisions: foraging theory for high-grading in a trawl fishery. *Behavioral Ecology* **6**, 146–154.

Gillis, D.M., Peterman, R.M., and Pikitch, E.K. (1995b) Implications of trip regulations for high-grading: a model of the behavior of fishermen. *Canadian Journal of Fisheries and Aquatic Sciences* **52**, 402–415.

Gilpin, M., and I. Hanski (1991) *Metapopulation dynamics.* Academic Press, New York.

Giske, J., Huse, G., and Fiksen, Ø. (1998) Modelling spatial dynamics of fish. *Reviews in Fish Biology and Fisheries* 8, 57–91.

Godfray, H.C.J. (1994) *Parasitoids.* Princeton University Press, Princeton, N.J.

Godfray, H.C.J., Partridge, L., and Harvey, P.H. (1991) Clutch size. *Annual Review of Ecology and Systematics* 22, 409–429.

Goldberg, D.E. (1989) *Genetic algorithms in search, optimization, and machine learning.* Addison-Wesley, Reading, Mass.

Gosler, A.G. (1987) Some aspects of bill morphology in relation to ecology in the great tit *Parus major.* D. Phil. Thesis, University of Oxford.

Gosler, A.G., Greenwood, J.J.D., and Perrins, C. (1995) Predation risk and the cost of being fat. *Nature* (London) 377, 621–623.

Gotelli, N. (1995) *A primer of ecology.* Sinauer Associates, Sunderland, Mass.

Gotelli, N.J., and Kelley, W.G. (1993) A general model of metapopulation dynamics. *Oikos* 68, 36–44.

Gould, S.J., and Lewontin, R.C. (1979) The spandrels of San Marco and the Panglossian paradigm: a critique of the adaptationist programme. *Proceedings of the Royal Society of London* B 205, 581–598.

Grafen, A. (1999) Formal Darwinism, the individual-as-maximizing agent analogy and bet-hedging. *Proceedings of the Royal Society of London* B 266, 799–803.

Gross, K. L. (1981) Predictions of fate from rosette size in four biennial plant species: *Verbascum thapsus, Oenothera biennis, Daucus carota,* and *Tragopogon dubius. Oecologia* 48, 209–213.

Gross, M.R. (1987) Evolution of diadromy in fish. *American Fisheries Society Symposium.* 1, 14–25.

Gubbary, S. (1995) *Marine protected areas: Principles and techniques for management.* Chapman and Hall, London.

Haftorn, S. (1956) Contribution to the food biology of tits, especially about storing of surplus. Part II. The Coal Tit (*Parus a. ater* L.). *K. Norske Vidensk. Selsk.* Skr. 2 1–52.

Haftorn, S. (1992) The diurnal body weight cycle in titmice *Parus* spp. *Ornis Scandinavica* 23, 435–443.

Hairston, N.G. (1989) *Ecological experiments: purpose, design and execution.* Cambridge University Press, New York.

Hamilton, W.D. (1964) The genetical evolution of social behavior. *Journal of Theoretical Biology* 7, 1–52.

Hanski, I. (1989) Metapopulation dynamics: does it help to have more of the same? *Trends in Ecology and Evolution* 4, 113–114.

Hanksi, I. (1994) A practical model of metapopulation dynamics. *Journal of Animal Ecology* 63, 151–162.

Hanski, I., Moilanen, A., Pakkala, T., and Kuussaari, M. (1996) The quantitative incidence function model and persistence of an endangered butterfly metapopulation. *Conservation Biology* 10, 578–590.

Harris, T.E. (1963) *The theory of branching processes.* Springer-Verlag, Berlin.

Hart, P.J.B. (1994) Theoretical reflections on the growth of three-spined stickleback morphs from inland lakes. *Journal of Fish Biology* 45 (Supplement A), 27–40.

Hart, P.J.B. (1997) Controlling illegal fishing in closed areas: the case of mackerel off Norway. In D.A. Hancock, D.C. Smith, A. Grant, and J.P. Beumer (eds.), *Developing and sustaining world fisheries*. Proceedings of the 2nd World Fisheries Congress. CSIRO, Collingwood, Australia, pp. 411–414.

Hart, P.J.B., and Gill, A.B. (1992a) Choosing prey size: a comparison of static and dynamic foraging models for predicting prey choice by fish. *Marine Behavior and Physiology* **22**, 93–106.

Hart, P.J.B., and Gill, A.B. (1992b) Constraints on prey size selection by the three-spined stickleback: energy requirements and the capacity and fullness of the gut. *Journal of Fish Biology* **40**, 205–218.

Hart, P.J.B., and Ison, S. (1991) The influence of prey size and abundance, and individual phenotype on prey choice by the three-spined stickleback, *Gasterosteus aculeatus* L. *Journal of Fish Biology* **38**, 359–372.

Harvell, C.D. (1990) The ecology and evolution of inducible defenses. *Quarterly Review of Biology* **65**, 323–340.

Harvey, P.H., and Pagel, M.D. (1991) *The comparative method in evolutionary biology*. Oxford University Press, New York.

Hassell, M.P. (1978) *The dynamics of arthropod predator-prey systems*. Princeton University Press, Princeton, N.J.

Hedenström, A., and Alerstam, T. (1995) Optimal flight speed of birds. *Philosophical Transactions of the Royal Society of London* B **348**, 471–487.

Heimpel, G.E., and Collier, T.R. (1996) The evolution of host-feeding behaviour in insect parasitoids. *Biological Reviews* **71**, 373–400.

Heimpel, G.E., and Rosenheim, J.A. (1995) Dynamic host feeding by the parasitoid *Aphytis melinus*: the balance between current and future reproduction. *Journal of Animal Ecology* **64**, 153–167.

Heimpel, G.E, Rosenheim, J., and Adams, J.M. (1994) Behavioral ecology of host feeding in Aphytis parasitoids. *Journal of Agricultural Sciences* Supplement **16**, 101–115.

Heimpel, G.E., Mangel, M., and Rosenheim, J.A. (1998) Effects of time limitation and egg limitation on lifetime reproductive success of a parasitoid in the field. *American Naturalist* **152**, 273–289.

Heimpel, G.E., Rosenheim, J.A., and Mangel, M. (1996) Egg limitation, host quality, and dynamic behavior by a parasitoid in the field. *Ecology* **77**, 2410–2420.

Higgins, P.J., and Talbot, C. (1985) Growth and feeding in juvenile Atlantic Salmon (*Salmo salar* L.). In C.B. Cowey, A.M. Mackie and J.G. Bell (eds.) *Nutrition and feeding in fish*. Academic Press, London, pp. 243-263.

Hilborn R., and Mangel, M. (1997) *The ecological detective: confronting models with data*. Princeton University Press, Princeton, N.J.

Hill, K., and Hurtado, M. (1991) The evolution of premature reproductive senescence and menopause in human females: an evaluation of the "grandmother" hypothesis. *Human Nature* **2**, 313–350.

Hitchcock, C.L., and Houston, A.I. (1994) The value of a hoard: not just energy. *Behavioral Ecology* **5**, 202–205.

Hitchcock, C.L., and Sherry, D.F. (1990) Long-term memory for cache sites in the black-capped chickadee. *Animal Behavior* **40**, 701–712.

Hoffmeister, T.S., and Roitberg, B.D. (1998) Evolution in signal persistence under predator exploitation. *EcoScience* **5**, 312–320.

Hogstad, O. (1988) Social rank and antipredator behavior of willow tits *Parus montanus* in winter flocks. *Ibis* **130**, 45–56.

Hogstad, O. (1989) Subordination in mixed-age bird flocks—a removal study. *Ibis* **131**, 128–134.

Holland, J. (1975) *Adaptation in natural and artificial systems*. University of Michigan Press, Ann Arbor, Mich.

Holling, C.S. (1959) Some characteristics of simple types of predation and parasitism. *Canadian Entomologist* **91**, 385–398.

Houston, A.I., and McNamara, J.M. (1987) Singing to attract a mate: a stochastic dynamic game. *Journal of Theoretical Biology* **129**, 57–68.

Houston, A.I., and McNamara, J.M. (1988) Fighting for food: a dynamic version of the Hawk-Dove game. *Evolutionary Ecology* **2**, 51–64.

Houston, A.I., and McNamara, J.M. (1993) A theoretical investigation of the fat reserves and mortality levels of small birds in winter. *Ornis Scandinavica* **24**, 205–219.

Houston, A.I., and McNamara, J.M. (1999) *Models of adaptive behaviour: An approach based on state*. Cambridge University Press, Cambridge, U.K.

Howell, N. (1979) *Demography of the Dobe !Kung*. Academic Press, New York.

Howson, C., and Urbach, P. (1989) *Scientific reasoning: The Bayesian approach*. Open Court Publishing, La Salle, Ill.

Howson, C., and Urbach, P. (1991) Bayesian reasoning in science. *Nature* **350**, 371–374.

Hugie, D.M., and Dill, L.M. (1994) Fish and game: a game-theoretic approach to habitat selection by predators and prey. *Journal of Fish Biology* **45** (Supplement A), 151–169.

Huse, G., and Giske, J. (1998) Ecology in the Mare Pentium: An individual-based spatio-temporal model for fish with adapted behaviour. *Fisheries Research* **37**, 163–168.

Iverson, G.C., Warnock, S.E., Butler, R.W., Bishop, M.A., and Warnock, N. (1996) Spring migration of Western Sandpipers along the Pacific Coast of North America: a telemetry study. *Condor* **98**, 10–21.

Jansson, C. (1982) The year round diets of the Willow Tit (*Parus montanus*) Conrad and the Crested Tit (*P. cristatus*) L. Ph. D. Thesis, University of Göteborg.

Jervis, M.A., and Kidd, N.A.C. (1986) Host-feeding strategies in hymenopteran parasitoids. *Biological Reviews* **61**, 395–434.

Jervis, M.A., Kidd, N.A.C., Fitton, M.G., Huddleston, T., and Dawah, H.A. (1993) Flower-visiting by hymenopteran parasitoids. *Journal of Natural History* **27**, 67–105.

Jobling, M., and Miglavs, I. (1993) The size of lipid depots—A factor contributing to the control of food intake in arctic charr, *Salvelinus alpinus*. *Journal of Fish Biology* **43**, 487–489.

Johnson, E.A. (1992) *Fire and vegetation dynamics: Studies from the North American boreal forest*. Cambridge University Press, New York.

Kachi, N. (1990) Evolution of size-dependent reproduction in biennial plants: a demographic approach. In S. Kawano (ed.), *Biological approaches and evolutionary trends in plants*. Academic Press, London, pp. 367–385.

Kachi, N. and Hirose, T. (1985) Population-dynamics of *Oenothera glazioviana* in a sand-dune system with special reference to the adaptive significance of size-dependent reproduction. *Journal of Ecology* **73**, 887–901.

Karlin, S., and Taylor, H.M. (1975) *A first course in stochastic processes*. Houghton Mifflin, Boston, Mass.

Kelly, E.J., and Kennedy, P.L. (1993). A dynamic state variable model of mate desertion in Cooper's hawks. *Ecology* **74**, 351–366.

Kisdi, É., and Meszéna, G. (1993) Density dependent life history evolution in fluctuating environments. In J. Yoshimura and C.W. Clark (eds.), *Adaptation in stochastic environments*. Lecture Notes in Biomathematics, Vol. 98. Springer-Verlag, Berlin, pp. 26–62.

Klinkhamer, P.G.L, de Jong, T.J., and Meelis, E. (1987) Delay of flowering in the 'biennial' *Cirsium vulgare*: size effects and devernalization. *Oikos* **49**, 303–308.

Klinkhamer, P. G. L., Meelis, E., de Jong, T.J., and Weiner, J. (1992) On the analysis of size-dependent reproductive output in plants. *Functional Ecology* **6**, 308–316.

Klomp, H., and Teernik, B.J. (1967) The significance of oviposition rates in the egg parasite, *Trichogramma embryophaugm* HTG. *Netherlands Journal of Zoology* **17**, 350–375.

Krebs, J.R., and Davies, N.B., Eds. (1993) *Behavioral ecology*, 3rd edition. Sinauer Associates, Sunderland, Mass.

Krebs, J.R., Kacelnik, A., and Taylor, P. (1978) Test of optimal sampling by foraging great tits. *Nature* **275**, 27–31.

Lack, D. (1946) Clutch and brood size in the robin. *British Birds* **39**, 98–109, 130–135.

Lack, D. (1947) The significance of clutch size. *Ibis* **89**, 302–352.

Lack D. (1948a) The significance of clutch size. *Ibis* **90**, 24–45.

Lack, D. (1948b) Further notes on clutch and brood size in the robin. *British Birds* **41**, 98–104, 130–137.

Lauck, T., Clark, C.W., Mangel, M., and Munro, G.R. (1998) Implementing the precautionary principle in fisheries management through marine reserves. *Ecological Applications* **8** (Supplement), S72–S78.

Law, R. (1979) Ecological determinants in the evolution of life histories. In R.M. Anderson, B.D. Turner, and L.R. Taylor, (eds.) *Population dynamics. The 20th symposium of the British Ecological Society*. Blackwell Scientific Publications, Oxford, U.K, pp. 81–103.

Leader-Williams, N., and Milner-Gulland, E.J. (1993) Policies for the enforcement of wildlife laws: the balance between detection and penalties in Luangwa Valley, Zambia. *Conservation Biology* **7**, 611-617.

Lee, R.B. (1972) The !Kung Bushmen of Botswana. In M.G. Bicchievi, (ed.) *Hunters and gatherers today*. Holt, Rinehart & Winston, New York.

Lee, R.B. (1979) *The !kung San. Men, women, and work in a foraging society*. Cambridge University Press, Cambridge, Mass.

Leslie, P.H. (1945) On the use of matrices in certain population mathematics. *Biometrika* **33**, 183–212.

Levin, S.A. (1992) The problem of pattern and scale in ecology. *Ecology* **73**, 1943–1967.

Levins, R. (1969) Some demographic and genetic consequences of environmental heterogeneity for biological control. *Bulletin of the Entomological Society of America* **15**, 237–240.

Levins, R. (1970) Extinction. In M. Gerstenhaber (ed.) *Some mathematical questions in biology*. American Mathematical Society, Providence, R.I., pp. 75–107.

Lewis, W.J., and Takasu, K. (1990) Use of learned odours by a parasitic wasp in accordance with host and food needs. *Nature* **348**, 635–636.

Lichtenbelt, W.D. v. M. (1993) Optimal foraging of a herbivorous lizard, the green iguana in a seasonal environment. *Oecologia* **95**, 246–256.

Liechti, F. (1995) Modelling optimal heading and airspeed of migrating birds in relation to energy expenditure and wind influence. *Journal of Avian Biology* **26**, 330–336.

Lima, S.L. (1986) Predation risk and unpredictable feeding conditions: determinants of body mass in birds. *Ecology* **67**, 377–385.

Lima, S.L. (1998) Stress and decision making under the risk of predation: recent developments from behavioral, reproductive, and ecological perspectives. *Advances in the Study of Behavior* **27**, 215–290.

Lindstrom, Å. (1991) Maximum fat deposition rates in migrating birds. *Ornis Scandinavica* **22**, 12–19.

Loehle, C., and Li, B. (1996) Habitat destruction and the extinction debt revisited. *Ecological Applications* **6**, 784–789.

Lomnicki, A. (1988) *Population ecology of individuals*. Princeton University Press, Princeton, N.J.

Louda, S.M., and Potvin, M.A. (1995) Effect of inflorescence-feeding insects on the demography and lifetime fitness of a native plant. *Ecology* **76**, 229–245.

Lucas, J.R., and Walter, L.R. (1991) When should chickadees hoard food? Theory and experimental results. *Animal Behaviour* **41**, 579–601.

Lucas, J.R., Creel, S.R., and Waser, P.M. (1995) Dynamic optimization and cooperative breeding: an evaluation of future fitness effects. In N.G. Solomon and J.A. French, (eds.). *Cooperative breeding in mammals*. Cambridge University Press, Cambridge, U.K, pp. 171–198.

Lucas, J.R., Howard, R.D., and Palma, J.G. (1996) Callers and satellites: chorus behaviour in anurans as a stochastic dynamic game. *Animal Behavior* **51**, 501–518.

MacArthur, R.H. (1962) Some generalized theorems on natural selection. *Proceedings of the National Academy of Sciences USA* **48**, 1893–1897.

MacArthur, R.H., and Pianka, E.R. (1966) On the optimal use of a patchy environment. *American Naturalist* **100**, 603–609.

MacArthur, R.H., and Wilson, E.O. (1967) *The theory of island biogeography*. Princeton University Press, Princeton, N.J.

Mace, R. (1995) Why do we do what we do? *Trends in Ecology and Evolution* **10**, 4–5.

Mace, R. (1996) When to have another baby: a dynamic model of reproductive decision-making and evidence from Gabbra pastoralists. *Ethology and Sociobiology* **17**, 263–273.

Mallows, C.L. (1973) Some comments on C_p. *Technometrics* **15**, 661–675.

Mangel, M. (1985) *Decision and control in uncertain resource systems*. Academic Press, New York.

Mangel, M. (1986) Solution of functional difference equations from behavioral theory. *Journal of Mathematical Biology* **24**, 557–567.

Mangel, M. (1987) Oviposition site selection and clutch size in insects. *Journal of Mathematical Biology* **25**, 1–22.

Mangel, M. (1989) Evolution of host selection in parasitoids: Does the state of the parasitoid matter? *American Naturalist* **133**, 688–705.

Mangel, M. (1990a) A dynamic habitat selection game. *Mathematical Biosciences* **100**, 241–248.

Mangel, M. (1990b) Evolutionary and neural network models of behavior. *Journal of Mathematical Biology* **28**, 237–256.

Mangel, M. (1990c). Dynamic information in uncertain and changing worlds. *Journal of Theoretical Biology* **146**, 317–332.

Mangel, M. (1991) Adaptive walks on behavioral landscapes and the evolution of optimal behavior by natural selection. *Evolutionary Ecology* **5**, 30–39.

Mangel, M. (1992) Rate maximizing and state variable theories of diet selection. *Bulletin of Mathematical Biology* **54**, 413–422.

Mangel, M. (1993) Effects of high seas driftnet fisheries on the Northern Right Whale Dolphin (*Lissodelphus Borealis*). *Ecological Applications* **3**, 221–229.

Mangel, M. (1994) Climate change and salmonid life history variation. *Deep Sea Research. Part II. Tropical Studies in Oceanography* **41**, 75–106.

Mangel, M., and Clark, C.W. (1983) Uncertainty, search, and information in fisheries. *Journal du Conseil International pour l'Exploration de la Mer* **41**, 93–103.

Mangel, M., and Clark, C.W. (1986) Towards a unified foraging theory. *Ecology* **67**, 1127–1138.

Mangel, M., and Clark, C.W. (1988) *Dynamic modeling in behavioral ecology*. Princeton University Press, Princeton, N.J.

Mangel, M., and Roitberg, B.D. (1988) On the evolutionary ecology of marking pheromones. *Evolutionary Ecology* **2**, 289–315.

Mangel, M., and Roitberg, B.D. (1989) Dynamic information and host acceptance by a tephritid fruit fly. *Ecological Entomology* **14**, 181–189.

Mangel, M., and Roitberg, B.D. (1992) Behavioral stabilization of host-parasite population dynamics. *Theoretical Population Biology* **42**, 308–320.

Mangel, M., and Roitberg, B.D. (1993) Larval life-styles and oviposition behavior of parasites and grazers. *Evolutionary Ecology* **7**, 401–406.

Mangel, M., Fiksen, O., and Giske, J. (1998) Logical, statistical and theoretical models in natural resource management and research. In A. Franklin and T. Schenk (eds.), *How to practice safe modeling*. Island Press, Boulder, Colo.

Martz, H.F., and Waller, R.A. (1982) *Bayesian reliability analysis*. John Wiley & Sons, New York.

May, R.M. (1978) Host-parasitoid systems in patchy environments: a phenomenological model. *Journal of Animal Ecology* **47**, 833–843.

Mayer, P.J. (1982) Evolutionary advantage of the menopause. *Human Ecology* **10**, 477–494.

Maynard Smith, J. (1982) *Evolution and the theory of games*. Cambridge University Press, Cambridge, U.K.

Mayr, E. (1976) *Evolution and the diversity of life. Selected essays*. Belknap Press, Harvard University, Cambridge Mass.

McGregor, R. (1997) Host-feeding and oviposition by parasitoids on hosts of different fitness value: influences of egg load and encounter rate. *Journal of Insect Behavior* **10**, 451–462.

McNamara, J.M. (1992) Optimal life histories: a generalisation of the Perron-Frobenius Theorem. *Theoretical Population Biology* **40**, 230–245.

McNamara, J.M. (1997) Optimal life histories for structured populations in fluctuating environments. *Theoretical Population Biology* **51**, 94–108.

McNamara, J.M., and Houston, A.I. (1982) Short term behavior and lifetime fitness. In D.J. McFarland (ed.), *Functional ontogeny*. Pitman, London, pp. 60–87.

McNamara, J.M., and Houston, A.I. (1986) The common currency for behavioral decisions. *American Naturalist* **127**, 358–378.

McNamara, J.M., and Houston, A.I. (1987) Starvation and predation as factors limiting population size. *Ecology* **68**, 1515–1519.

McNamara, J.M., and Houston, A.I. (1990) State-dependent ideal free distributions. *Evolutionary Ecology* **4**, 298–311.

McNamara, J.M., and Houston, A.I. (1996) State-dependent life histories. *Nature* **380**, 215–221.

McNamara, J.M., Houston, A.I., and Krebs, J.R. (1990) Why hoard? The economics of food storing in tits, *Parus* spp. *Behavioral Ecology* **1**, 12–23.

McNamara, J.M., Houston, A.I., and Lima, S.L. (1994) Foraging routines of small birds in winter: a theoretical investigation. *Journal of Avian Biology* **25**, 287–302.

McNamara, J.M., Webb, J.N., and Collins, E.J. (1995) Dynamic optimization in fluctuating environments. *Proceedings of the Royal Society of London* B **261**, 279–284.

McNamara, J.M., Webb, J.N., Collins, E.J., Székely, T., and Houston, A.I. (1997) A general technique for computing evolutionarily stable strategies based on errors in decision making. *Journal of Theoretical Biology* **189**, 211–225.

Messing, R.H., Klugness, L.M., and Purcell, M.F. (1994) Short-range dispersal of mass-reared *Diachasmimorpha longicaudata* and *D. tyroni* (Hymenoptera: Braconidae), parasitoids of tephritid fruit flies. *Journal of Economic Entomology* **87**, 975–985.

Metcalfe, N.B., and Thorpe, J.E. (1992) Anorexia and defended energy levels in over-wintering juvenile salmon. *Journal of Animal Ecology* **61**, 175–181.

Metcalfe, N.B., and Ure, S.E. (1995) Diurnal variation in flight performance and hence potential predation risk in small birds. *Proceedings of the Royal Society of London* B **261**, 395–400.

Metcalfe, N.B., Huntingford, F.A., and Thorpe, J.E. (1986) Seasonal changes in feeding motivation of juvenile Atlantic salmon (*Salmo salar*). *Canadian Journal of Zoology* **64**, 2439–2446.

Metcalfe, N.B., Huntingford, F.A., and Thorpe, J.E. (1988) Feeding intensity, growth rates, and the establishment of life-history patterns in juvenile Atlantic salmon *Salmo salar*. *Journal of Animal Ecology* **57**, 463–474.

Metcalfe, N.B., Taylor, A.C., and Thorpe, J.E. (1995) Metabolic rate, social status, and life-history strategies in Atlantic salmon. *Animal Behaviour* **49**, 431–436.

Metz, J.A.J., Nisbet, R.M., and Geritz, S.A.H. (1992). How should we define 'fitness' for general ecological scenarios? *Trends in Ecology and Evolution* **7**, 198–202.

Meyhöfer, R., Casas, J., and Dorn, S. (1994) Host location by a parasitoid using leafminer vibrations: characterizing the vibrational signals produced by the leafmining host. *Physiological Entomology* **19**, 349–359.

Milner-Gulland, E.J., and Leader-Williams, N. (1992) A model of incentives for the illegal exploitation of black rhinos and elephants: poaching pays in Luangwa Valley, Zambia. *Journal of Applied Ecology* **29**, 388-401.

Minkenberg, O.P.J.M., Tatar, M., and Rosenheim, J.A. (1992) Egg load as a major source of variability in insect oviposition behavior. *Oikos* **65**, 134–142.

Mittelbach, G.G. (1981) Foraging efficiency and body size: a study of optimal diet and habitat use by bluegills. *Ecology* **62**, 1370–1386.

Monaghan, P., and Nager, R.G. (1997) Why don't birds lay more eggs? *Trends in Ecology and Evolution* **12**, 270–274.

Moore, A.D. (1990) The semi-Markov process: a useful tool in the analysis of vegetation dynamics for management. *Journal of Environmental Management* **30**, 111–130.

Morris, W.F. (1993) Predicting the consequences of plant spacing and biased movement for pollen dispersal by honey bees. *Ecology* **74**, 493–500.

Morris, W.F., Mangel, M., and Adler, F.R. (1995) Mechanisms of pollen deposition by insect pollinators. *Evolutionary Ecology* **9**, 304–317.

Murdoch, W.W. (1994) Population regulation in theory and practice. *Ecology* **75**, 271–287.

Murdoch, W.W., and Stewart-Oaten, A. (1989) Aggregation by parasitoids and predators. Effects on equilibrium and stability. *American Naturalist* **134**, 288–310.

Murray, J.D. (1990) *Mathematical biology*. Springer Verlag, New York.

Mylius, S.D., and Diekmann, O. (1995) On evolutionarily stable life histories, optimization and the need to be specific about density dependence. *Oikos* **74**, 218–224.

Nakamura, S. (1997) Clutch size regulation and host discrimination of the parasitoid fly, *Exorista japonica* (Diptera: Tachinidae). *Applied Entomology and Zoology* **32**, 283–291.

Nettleship, D.N., and Birkhead, T.R. (1986) *Atlantic alcidae*. Academic Press, London.

Nicieza, A.G., Brana, F., and Toledo, M.M. (1991) Development of length-bimodality and smolting in wild stocks of Atlantic salmon, *Salmo salar*, under different growth conditions. *Journal of Fish Biology* **38**, 509–523.

Norberg, U.M. (1990) *Vertebrate flight*. Springer, Berlin.

Norberg, U.M. (1996) Energetics of flight. In C. Carey (ed.) *Avian energetics and nutritional ecology*. Chapman and Hall, New York, NY, pp. 199–249.

Nuechterlein, G.L. (1981) Courtship behavior and reproductive isolation between western grebe color morphs. *The Auk* **98**, 335–349.

Nuechterlein, G.L., and Buitron, D. (1998) Interspecific mate choice by late-courting male western grebes. *Behavioral Ecology* **9**, 313–321.

Nuechterlein, G.L., and Storer, R.W. (1982) The pair-formation displays of the western grebe. *The Condor* **84**, 350–369.

Nuechterlein, G.L., and Storer, R.W. (1989) Mate feeding by western and Clark's grebes. *The Condor* **91**, 37–42.

Okubo, A. (1980) *Diffusion and ecological problems: Mathematical models*. Springer-Verlag, New York.

Orzack, S.H. (1993) Life history evolution and population dynamics in variable environments: Some insights from stochastic demography. In J. Yoshimura and C.W. Clark (eds.), *Adaptation in stochastic environments*. Lecture Notes in Biomathematics, Vol. 98. Springer-Verlag, Berlin, pp. 63–104.

Oster, G.F., and Wilson, E.O. (1978) *Caste and ecology in the social insects*. Princeton University Press, Princeton, N.J.

Pacala, S.W., Hassell, M.P., and May, R.M. (1990) Host-parasitoid associations in patchy environments. *Nature* **344**, 150–153.

Papaj, D.R., and Lewis, A.C., Eds. (1993) *Insect learning: Ecological and evolutionary perspectives*. Chapman and Hall, New York.

Pennycuick, C.J. (1975) Mechanics of flight. In D.S. Farner and J.R. King (eds.), *Avian Biology*, Vol. V. Academic Press, London, pp. 1–75.

Pennycuick, C.J. (1989) *Bird flight performance*. Oxford University Press, Oxford.

Pentelow, F.T.K., Southgate, B.A., and Bassindale, R. (1933) The relation between size, age, and time of migration of salmon and sea trout smolts in the river Tees. *Fisheries Investigations* Series 1 **3**, 3–14.

Peters, C.S., Mangel, M., and Costantino, R.F. (1989) Stationary distribution of population size in Tribolium. *Bulletin of Mathematical Biology* **51**, 625–638.

Pettersson, L.B., and Brönmark, C. (1993) Trading off safety against food: state dependent habitat choice and foraging in crucian carp. *Oecologia* **95**, 353–357.

Piersma, T. (1994) Close to the edge: energetic bottlenecks and the evolution of migratory pathways in Knots. Doctoral Dissertation, Rijksuniversiteit Groningen, Netherlands.

Piersma, T., and Gill, R.E., Jr. (1998) Guts don't fly: small digestive organs in obese Bar-tailed Godwits. *The Auk* **115**, 196–203.

Plomin, R., DeFries, J.C., and McClearn, G.E. (1990) *Behavioral genetics. A primer*. 2nd edition. W.H. Freeman, San Francisco, CA.

Possingham, H. (1996) Decision theory and biodiversity management: How to manage a metapopulation. In R.B. Floyd, A.W. Sheppard and P.J. De Barro (eds.) *Frontiers of population ecology*. CSIRO, Melbourne, Australia, pp. 391–398.

Possingham, H. (1997) State-dependent decision analysis for conservation biology. In S.T.A. Pickett, R.S. Ostfeld, M. Schachak and G.E. Likens (eds.), *The ecological basis of conservation: Heterogeneity, ecosystems and biodiversity*. Chapman and Hall, New York, NY, pp. 298–304.

Possingham, H., and Tuck, G. (1997) Application of stochastic dynamic programming to optimal fire management of a spatially structured threatened species. In A. D. McDonald and M. McAleer (eds.), *Proceedings International Congress on Modelling and Simulation*, Modelling and Simulation Society of Australia Inc., Vol.2, pp. 813–817.

Possingham, H., Day, J. Goldfinch, M., and Salzborn, F. (1993) The mathematics of designing a network of protected areas for conservation. In D. Sutton, F. Cousings, and C. Pearce (eds.) *Proceedings of the 12th Australian Operations Research Conference*, Adelaide University. University of Adelaide Press, Adelaide, Australia, pp. 536–545.

Press, W.H., Flannery, B.P., Teukolsky, S.A., and Vetterling, W.T. (1986) *Numerical recipes. The art of scientific computing*. Cambridge University Press, New York.

Price, K. (1994) The behavioral ecology of begging in Yellow-headed Blackbird nestlings. Ph.D. Thesis, Simon Fraser University, Burnaby, B.C., Canada.

Price, P.V. (1980) *The evolutionary biology of parasites*. Princeton University Press, Princeton, N.J.

Pyke, G.H., Saillard, R., and Smith, J. (1995) Abundance of eastern bristlebirds in relation to habitat and fire history. *Emu* **95**, 106–110.

Rayner, J.M.V. (1988) Form and function in avian flight. *Current Ornithology* **5**, 1–66.

Rees, M., and Crawley, M.J. (1989) Growth, reproduction and population-dynamics. *Functional Ecology* **3**, 645–653.

Rees, M., Sheppard, A., Briese, D., and Mangel, M. (1999) Evolution of size dependent flowering in *Onopordum illyricum*. *American Naturalist* (in press).

Reinartz, J.A. (1984) Life-history variation of common mullein(*Verbascum thapsus*). 1. Latitudinal differences in population-dynamics and timing of reproduction. *Journal of Ecology* **72**, 897–912.

Rhymer, J.M., and Simberloff, D. (1996) Extinction by hybridization and introgression. *Annual Review of Ecology and Systematics* **27**, 83–109.

Risch, T.S., Dobson, F.S., and Murie, J.O. (1995) Is mean litter size the most productive? A test in Columbian ground squirrels. *Ecology* **76**, 1643–1654.

Rivero-Lynch, A.P., and Godfray, H.C.J. (1997) The dynamics of egg production, oviposition and resorption in a parasitoid wasp. *Functional Ecology* **11**, 184–188.

Roberts, C.M. (1997) Ecological advice for the global fisheries crisis. *Trends in Ecology and Evolution* **12**, 35–38.

Roberts, R.C. (1979) The evolution of avian food-storing behavior. *American Naturalist* **114**, 418–438.

Robertson, I.C., Robertson, W.G., and Roitberg, B.D. (1998) A model of mutual tolerance and the origin of communal association between unrelated females. *Journal of Insect Behavior* **11**, 265–286.

Roff, D.A. (1992) *The evolution of life histories: Theory and analysis*. Chapman and Hall, New York.

Roitberg, B.D. 1990. Optimistic and pessimistic fruit flies: evaluating fitness consqeunces of estimation errors. *Behaviour* **114**, 65–82.

Roitberg, B.D., and Mangel, M. (1993) Parent-offspring conflict and life-history consequences in herbivorous insects. *American Naturalist* **142**, 443–456.

Roitberg, B.D., Mangel, M., Lalonde, R.G., Roitberg, C.A., van Alphen, J.J.M., and Vet, L. (1992) Seasonal dynamic shifts in patch exploitation by parasitic wasps. *Behavioral Ecology* **3**, 156–165.

Roitberg, B.D., Sircom, J., Roitberg, C.A., van Alphen, J.J.M., and Mangel, M. (1993) Life expectancy and reproduction. *Nature* **364**, 108.

Rosenheim, J.A., and Heimpel, G.E. (1994) Sources of intraspecific variation in oviposition and host-feeding behavior. In D. Rosen (ed.), *Advances in the study of Aphytis*. Intercept Press, Andover, U.K, pp. 41–78.

Rosenheim, J.A., and Rosen, D. (1991) Foraging and oviposition decisions in the parasitoid *Aphytis lingnanensis*: distinguishing the influences of egg load and experience. *Journal of Animal Ecology* **60**, 873–893.

Rosenheim, J.A., and Rosen, D. (1992) Influence of egg load and host size on host-feeding behaviour of the parasitoid *Aphytis lingnanensis*. *Ecological Entomology* **17**, 263–272.

Rosenzweig, M.L. (1995) *Species diversity in space and time*. Cambridge University Press, New York.

Rosenzweig, M.L., and Abramsky, Z. (1997) Two gerbils of the Negev: a long-term investigation of optimal habitat selection and its consequences. *Evolutionary Ecology* **11**, 733–756.

Rosenzweig, M.L., and Clark, C.W. (1994) Island extinction rates from regular censuses. *Conservation Biology* **8**, 491–494.

Salvanes, A.G.V., and Hart, P.J.B. (1998) Individual variability in state-dependent feeding behaviour in three-spined sticklebacks. *Animal Behaviour* **55**, 1349–1359.

Samson, D.A., and Werk, K.S. (1986) Size-dependent effects in the analysis of reproductive effort. *American Naturalist* **127**, 667–680.

Sandlan, K.P. (1979) Host-feeding and its effects on the physiology and behaviour of the ichneumonid parasitoid, *Coccygomimus turionellae*. *Physiological Entomology* **4**, 383–392.

Schat, H., Ouborg, J., and De Wit, R. (1989) Life history and plant architecture: size-dependent reproductive allocation in annual and biennial *Centaurium* species. *Acta Botanica Neerlandica* **38**, 183–201.

Schlichting, C.D., and Pigliucci, M. (1998) *Phenotypic evolution. A reaction norm perspective.* Sinauer Associates, Sunderland, Mass.

Schwinning, S., and Rosenzweig, M.L. (1990) Periodic oscillations in an ideal-free predator-prey distribution. *Oikos* **59**, 85–91.

Seger, J., and Brockmann, J. (1987) What is bet-hedging? *Oxford Surveys in Evolutionary Biology* **4**, 182–211.

Shackell, N.L., and Willison, J.H.M., Eds. (1995) *Marine protected areas and sustainable fisheries.* Science and Management of Protected Areas Association, Acadia University, Wolfville, N.S., Canada.

Sherry, D.F. (1985) Food storage by birds and mammals. *Advances in the Study of Behavior* **15**, 153–188.

Simpson, A.L. (1993) Investigation of the factors influencing maturation in Atlantic salmon, *Salmo salar* L., Parr. Ph.D. Thesis, University of Glasgow, Scotland.

Simpson, A.L., Metcalfe, N.B., and Thorpe, J.E. (1992) A simple non-destructive biometric method for estimating fat levels in Atlantic salmon, *Salmo salar*. *Aquaculture and Fisheries Management* **23**, 23–29.

Smith, C.C., and Reichman, O.J. (1984) The evolution of food caching by birds and mammals. *Annual Review of Ecology and Systematics* **15**, 329–351.

Smith, T.D. (1994) *Scaling fisheries. The scieince of measuring the effects of fishing.* Cambridge University Press, Cambridge and New York.

Sokoloff, A. (1974) *The biology of Tribolium*, Vol. 2. Clarendon Press, Oxford, U.K.

Solomon, M.E. (1949) The natural control of animal populations. *Journal of Animal Ecology* **18**, 1–35.

Stearns, S.C. (1992) *The evolution of life histories.* Oxford University Press, Oxford, U.K.

Stephens, D.W. (1991). Change, regularity, and value in the evolution of animal learning. *Behavioral Ecology* **2**, 77–89.

Stephens, D.W., and Krebs, J.R. (1986) *Foraging theory.* Princeton University Press, Princeton, N.J.

Sumida, B.H., Houston, A.I., McNamara, J.M., and Hamilton, W.D. (1990) Genetic algorithms and evolution. *Journal of Theoretical Biology* **147**, 59–84.

Sutherland, W.J. (1991) Flying in the face of reason. *Nature* **353**, 211–212.

Sutherland, W.J. (1996) *From individual behavior to population ecology.* Oxford University Press, Oxford, U.K.

Taylor, A.D. (1991) Studying metapopulation effects in predator-prey systems. *Biological Journal of the Linnean Society* **42**, 305–323

Thorpe, J.E. (1977) Bimodal distribution of length of juvenile Atlantic salmon (*Salmo salar* L.) under artificial rearing conditions. *Journal of Fish Biology* **11**, 175–184.

Thorpe, J.E. (1987) Environmental regulation of growth patterns in juvenile Atlantic salmon. In R.C. Summerfelt and G.E. Hall (eds.), *Age and growth of fish.* Iowa State University Press, Ames, Iowa, pp. 463–474.

Thorpe, J.E., Metcalfe, N.B., and Huntingford, F.A. (1992) Behavioural influences on life-history variation in juvenile Atlantic salmon, *Salmo salar*. *Environmental Biology of Fishes* **33**, 331–340.

Thorpe, J.E., Morgan, R.I.G., Ottaway, E.M., and Miles, M.S. (1980) Time of divergence of growth groups between potential 1+ and 2+ smolts among sibling Atlantic salmon. *Journal of Fish Biology* **17**, 13–21.

Turchin, P., and Thoeny, W.T. (1993) Quantifying dispersal of southern pine beetles with mark-recapture experiments and a diffusion model. *Ecological Applications* **3**, 187–198.

Vadas, H.L., Burrows, M.T., and Hughes, R.N. (1994) Foraging strategies of dogwhelks, *Nucella lapillus* (L.): Interacting effects of age, diet and chemical cues to the threat of predation. *Oecologia* **100**, 439–450.

van der Meijden, E., and van der Waals-Kooi, R.E. (1979) The population ecology of *Senecio jacobaea* in a sand dune system. I. Reproductive strategy and the biennial habit. *Journal of Ecology* **67**, 131–153.

Vet, L.E.M., Wackers, F.I., and Dicke, M. (1991) How to hunt for hiding hosts: the reliability–detectability problem in foraging parasitoids. *Netherlands Journal of Zoology* **41**, 202–213.

Vet, L.E.M., Datema, A., van Welzen, K., and Snellen, H. (1993) Clutch size in a larval-pupal endoparasitoid 1. Variation across and within host species. *Oecologia* **95**, 410–415.

Vincent, A.C.J., and Hall, H.J. (1996) The threatened status of marine fishes. *Trends in Ecology and Evolution* **11**, 360–361.

Visser, M.E., and Rosenheim, J.A. (1998) The influence of competition between foragers on clutch size decisions in insect parasitoids. *Biological Control* **11**, 169–174.

Wäckers, F.L. (1994) The effect of food deprivation on the innate visual and olfactory preferences in the parasitoid *Cotesia rubecula*. *Journal of Insect Physiology* **40**, 641–649.

Wäckers, F.L., and Swaans, C.P.M. (1993) Finding floral nectar and honeydew in *Cotesia rubecula*: random or directed? *Proceedings Experimental and Applied Entomology* **4**, 67–72.

Wahlberg, N., Moilanen, A., and Hanski, I. (1996) Predicting the occurrence of endangered species in fragmented landscapes. *Science* **273**, 1536–1538.

Walde, S., and Murdoch, W.W. (1988) Spatial density dependence in parasitoids. *Annual Review of Entomology* **33**, 441–466.

Wallace, B. (1982) Phenotypic variation with respect to fitness: The basis for rank order selection. *Biological Journal of the Linnean Society* **17**, 269–274.

Walters, C.J. (1986) *Adaptive management of renewable resources*. Macmillan, New York.

Walters, C.J., and Hilborn, R. (1978) Ecological optimization and adaptive management. *Annual Review on Ecology and Systematics* **9**, 157–188.

Watt, K.E.F. (1968) *Ecology and resource management*. McGraw-Hill, New York.

Weber, T.P., and Houston, A.I. (1997) Flight costs, flight range and the stopover ecology of migrating birds. *Journal of Animal Ecology* **66**, 297–306.

Weber, T.P., Ens, B.J., and Houston, A.I. (1998) Optimal avian migration: a dynamic model of fuel stores and site use. *Evolutionary Ecology* **12**, 377–401.

Weber, T.P., Houston, A.I., and Ens, B.J. (1994) Optimal departure fat loads and stopover site use in avian migration: an analytical model. *Proceedings of the Royal Society of London* B **258**, 29–34.

Wellington, W.G. (1946) The effect of variations in atmospheric pressure upon insects. *Canadian Journal of Research.* Sec. D **24**, 51–58.

Werner, E.E., and Gilliam, J.F. (1984) The ontogenetic niche and species interactions in size-structured populations. *Annual Review of Ecology and Systematics* **15**, 393–425.

Werner, E.E., and Hall, D.J. (1974) Optimal foraging and size selection of prey by bluegill sunfish (*Lepomis macrochirus*). *Ecology* **55**, 1042–1052.

Werner, E.E., and Hall, D.J. (1988) Ontogenetic habitat shifts in bluegill: the foraging rate-predation risk trade-off. *Ecology* **69**, 1352–1366.

Werner, E.E., and Mittelbach, G.G. (1981) Optimal foraging theory: field tests of diet choice and habitat switching. *American Zoologist* **21**, 813–829.

Werner, E.E., Mittelbach, G.G., Hall, D.J., and Gilliam, J.F. (1983a) Experimental tests of optimal habitat use in fish: the role of relative habitat profitability. *Ecology* **64**, 1525–1539.

Werner, E.E., Hall, D.J., Mittelbach, G.G., and Gilliam, J.F. (1983b) An experimental test of the effects of predation risk on habitat use in fish. *Ecology* **64**, 1540–1548.

Werner, P.A. (1975) Predictions of fate from rosette size in teasel (*Dipsacus fullonum*) L. *Oecologia* **20**, 197–201.

Whelan, R.J. (1995) *The ecology of fire.* Cambridge University Press, New York.

Williams, B.K. (1996) Adaptive optimization of renewable natural resources: solution algorithms and a computer program. *Ecological Modelling* **93**, 101–111.

Williams, G.C. (1966) *Adaptation and natural selection: a critique of some current evolutionary thought.* Princeton University Press, Princeton, N.J.

Wilson, K. (1994) Evolution of clutch size in insects. II. A test of static optimality models using the beetle *Callosobruchus maculatus* (Coleoptera: Bruchidae) *Journal of Evolutionary Biology* **7**, 365–386.

Wilson, K., and Lessells, C.M. (1994) Evolution of clutch size in insects. I. A review of static optimality models. *Journal of Evolutionary Biology* **7**, 339–363.

Wilson, W.H. (1994) Western Sandpiper (*Calidris mauri*). In A. Poole and F. Gill (eds.) *The birds of North America*, No. 90. The Academy of Natural Sciences, Washington, D.C., and The American Ornithological Union.

Wynne-Edwards, V.C. (1962) *Animal dispersion in relation to social behaviour.* Hafner, New York.

Ydenberg, R.C. (1989) Growth mortality trade-offs and the evolution of juvenile life histories in the Alcidae. *Ecology* **70**, 1494–1506.

Ydenberg, R.C. (1998) Behavioral decisions about foraging and predator avoidance. In R. Dukas (ed.) *Cognitive Ecology.* University of Chicago Press, Chicago, IL, pp. 343–378.

Ydenberg, R.C., and Clark, C.W. (1989) Aerobiosis and anaerobiosis during diving by Western grebes: an optimal foraging approach. *Journal of Theoretical Biology* **139**, 437–449.

Ydenberg, R.C., Clark, C.W., and Harfenist, A. (1995) Intraspecific fledging mass variation in the Alcidae, with special reference to the seasonal fledging mass decline. *American Naturalist* **145**, 412–433.

Ydenberg, R.C., Welham, C.V.J., Schmid-Hempel, R.P., Schmid-Hempel, P., and Beauchamp, G. (1994) Time and energy constraints and the relationships between currencies in foraging theory. *Behavioral Ecology* **5**, 28–34.

Yoshimura, J., and Clark, C.W. (1993) *Adaptation in Stochastic Environments.* Springer-Verlag Lecture Notes in Biomathematics, 98. Springer-Verlag, Berlin.

Zwarts, L., Blomert, A.-M., Ens, B.J., Hupkes, R., and van Spanje, T.M. (1990) Why do waders reach high feeding densities on the intertidal flats of the Banc d'Arguin, Mauretania? *Ardea* **78**, 39–52.

Index